8/06

gen·e·sis

gen·e·sis

THE SCIENTIFIC QUEST FOR LIFE'S ORIGIN

Robert M. Hazen

Joseph Henry Press
Washington, DC

Joseph Henry Press • 500 Fifth Street, NW • Washington, DC 20001

The Joseph Henry Press, an imprint of the National Academies Press, was created with the goal of making books on science, technology, and health more widely available to professionals and the public. Joseph Henry was one of the founders of the National Academy of Sciences and a leader in early American science.

Library of Congress Cataloging-in-Publication Data

Hazen, Robert M., 1948-
  Genesis : the scientific quest for life's origin / by Robert M. Hazen.
      p. cm.
  Includes bibliographical references and index.
  ISBN 0-309-09432-1
  1. Life—Origin. I. Title.
  QH325.H39 2005
  576.8′3—dc22

                                2005012839

Cover design by Michele de la Menardiere.

Cover photo of microscopic vesicles courtesy of Robert M. Hazen and David W. Deamer. A key step in life's origin may have been the spontaneous assembly of these cell-like spheres of molecules.

Illustrations on pages 18, 145, 153, 159, 195, and 226-227 by Matthew Frey, Wood Ronsaville Harlin, Inc., Annapolis, Maryland. Copyright Wood Ronsaville Harlin, Inc.

Printed in the United States of America.

For Glenn, Steve, and Hat

Contents

Foreword

A central theme of this book is the concept of emergence. What do we mean when we say that something emerges? In common usage, a shadowy figure emerges from the dark, a submarine emerges from the sea, a plot emerges in a novel. But emergence has come to have a different meaning in scientific terminology. Researchers are increasingly beginning to use emergence to describe processes by which more complex systems arise from simpler systems, often in unpredictable fashion.

This use of the word "emergence" is in a sense the opposite of reductionism, the view that any phenomenon can be explained by understanding the parts of that system. Because reductionism has been such a powerful tool in the sciences, some scientists shy away from the concept of emergence, thinking it to be slightly weird. But there can be little doubt that the word itself is useful in referring to some of the most remarkable phenomena we observe in both nature and in the laboratory.

An example is the way that orderly arrangements of molecules can appear spontaneously. For instance, if we add soap molecules to water, at first there is nothing present but the expected clear solution of individual molecules dissolved in the water. But at a certain concentration, additional molecules no longer dissolve but instead begin to associate into small aggregates called vesicles. And as the concentration increases, the vesicles begin to grow into membranous layers of soap molecules that cloud the originally clear solution. Finally, if we blow air through a straw into the solution, much larger structures—soap bubbles—appear at the surface.

Such emergent phenomena—phenomena that exhibit self-organization—are common in our everyday experience. The laws of chemistry and physics permit certain kinds of molecules to self-assemble into aggregates that have surprising structures and properties. Sometimes the process is spontaneous, as in the formation of vesicles, but in other instances an input of energy is required to drive self-assembly. If we did not know by observation that soap molecules can self-assemble, we could not have predicted that vesicles would suddenly appear if we simply increased the concentration of soap molecules in solution. And even though we know that vesicles form, there is still no equation that can predict exactly what concentration of soap is required to form them.

Life on Earth arose through a sequence of many such emergent phenomena, which define the subject of this book. Imagine that we could somehow travel back in time to the prebiotic Earth, some 4 billion years ago. It is very hot—hotter than the hottest desert today. Asteroid-sized objects bombard the surface. Comets crash through the atmosphere—no oxygen yet, just a mixture of carbon dioxide and nitrogen—and add more water to a globe-spanning ocean. Landmasses are present, but they are volcanic islands resembling Hawaii or Iceland, rather than continents.

Imagine that we are standing on one such island, on a beach composed of black lava rocks, with tide pools containing clear seawater. We can scrutinize that water with a microscope, but there is nothing living to be seen in it, only a dilute solution of organic compounds and salts. If we could examine the mineral surfaces of the lava rocks, we would see that some of the organic compounds have formed a film adhering to the surface, while others have assembled into aggregates that disperse into the seawater.

Now imagine that we return 100 million years later. Not much has changed. The landmasses are still volcanic islands, meteorite impacts have dwindled, and it might be a little less hot. But, when we look at the tide pools, we see a cloudiness that was not apparent earlier, and the mineral surfaces are coated with a thin film of slime. When we examine the water and lava with our microscope, we discover immense numbers of bacteria swarming in multiple layers. Life has begun.

What happened in 100 million years that led to the origin of life?

This is a fundamental question of biology, and the answer will surely change the way we think about ourselves as well as our place in the universe, because if life could begin on Earth, it could begin by similar processes on Earth-like planets circling other stars throughout the universe. The origin of life is the most extraordinary example of an emergent phenomenon, and the process by which life began must involve the same kinds of intermolecular forces and self-assembly processes that cause soap to form membranous vesicles. The origin of life must also have in some way incorporated the reactions and products that occur when energy flows through a molecular system and drives it toward ever more complex systems with emergent properties.

This book explores the concept of emergence and the origin of life in a way that has never before been attempted. Science has thousands of investigators who pry away at highly focused aspects of the great questions, hardly aware of the vast unexplored problems spreading around them to the horizon. But each science has a few explorers— rare personalities willing to step back from the microscopic details, look toward the horizon, and gamble that patterns will emerge from their broader perspective. Robert Hazen is such an explorer, and this book is a journal of his explorations.

Genesis: The Scientific Quest for Life's Origin is a pretty amazing book. Many authors of popular science books are teachers and professors, and it is only natural that their books come across that way: as lectures—factual, conceptual, theoretical. Occasionally, an author is able to assemble facts, concepts, and theory in a creative way to produce a book that introduces a significant new paradigm. Darwin did that, and more recently E. O. Wilson and Stephen Jay Gould.

Hazen has taken a different approach, and a different set of words describes this book: It is personal, even intimate, filled with passion for the scientific enterprise. You will find facts, concepts, and theories here, too—but beyond that you will discover glimpses of scientists in action, chasing ideas in the lab and the field. You will find people struggling with experimental results, with interpretations, and with each other. You will find drama, which exists in the sciences as much as in any other human endeavor. And you will find cliffhangers: Will a future experiment show that Nick Platts' idea about primitive genetic polymers is correct? Or will it dash his hopes? Will new evidence permit a choice between the conflicting claims of Bill Schopf and Martin Brasier about the Apex Chert fossils?

Like Wilson and Gould, Hazen is a working scientist—a mineralogist who has broadened his field of research in order to tackle, with his colleagues at the Carnegie Institution's Geophysical Laboratory, some of the deepest problems of biology. Where did organic compounds come from to kick-start the life process on the early Earth? How did life become chiral (that is, "handed"), starting with mixtures of molecules that differ only in whether their structures rotate polarized light to the left or to the right? How did metabolic pathways arise from the interaction of organic molecules and mineral surfaces? A convincing answer to any one of these questions would be a capstone to a remarkable life in science. In this book you will learn how Hazen and other explorers are struggling to find those answers.

David Deamer
Santa Cruz, California
May 2005

Preface

How did life arise? Why are we here? For thousands of years humans
have longed for answers to these deeply resonant questions.

The Biblical account in the first chapter of Genesis, though rich in
poetic metaphor, hardly puts the origin question to rest. Barring di-
vine intervention, life must have emerged by a natural process—one
fully consistent with the laws of chemistry and physics. Scientists believe in a universe ordered by natural laws; they resort to the power of
observations, experiments, and theoretical reasoning to discover those
laws. The methods of science are unsuited to address the "why" of our
existence, but many of us feel driven to understand the nitty-gritty
chemical details of *how* life began.

Scientists surmise that life arose on the blasted, primitive Earth
from the most basic of raw materials: air, water, and rock. Life emerged
nearly 4 billion years ago by natural processes completely in accord
with the laws of chemistry and physics, yet details of that transforming
origin event pose mysteries as deep as any facing science. How did nonliving chemicals become alive?

It is possible, of course, that life arose through an improbable sequence of many chemical reactions. If so, then living worlds will be
rare in the universe and laboratory attempts to understand the origin
process will be doomed to frustration. An unlikely sequence of
unknown steps cannot be reproduced in any plausible experimental
program.

Alternatively, the universe may be organized in such a way that life emerges as an inevitable consequence of chemistry, given an appropriate environment and sufficient time. Starting with water, organic molecules, and a suitably protected energy-rich environment, life may be very likely to emerge from nonlife on any hospitable planet or moon. This scenario allows for fruitful systematic scientific study. If life is likely to arise whenever and wherever appropriate conditions occur, then scientists can hope to study life's origins in the lab through experiments that simulate those conducive conditions. Not surprisingly, most origin-of-life investigators favor the view that life is a cosmic imperative and that it is only a matter of time before we figure out how it happened. In this scenario, genesis occurs throughout the universe all the time.

Genesis: The Scientific Quest for Life's Origin attempts to portray this great adventure—the effort to deduce how life began on the ancient Earth. The epic history of life's chemical origins is woefully incomplete. Daunting gaps exist in our knowledge, and much of what we have learned is hotly debated and subject to conflicting interpretations. Consequently, this book is as much about the process of defining what we do not know as it is about recounting well-established data and concepts. One objective of the book is to describe our present, imperfect state of understanding—and to offer a conceptually simple scenario for life's chemical origins. This theory synthesizes two fundamental frontier efforts: the mind-expanding theoretical field of emergence and the astonishing experimental discoveries in prebiotic chemistry.

The science of emergence seeks to understand complex systems—systems that display novel collective behaviors that arise from the interactions of many simple components. From gravitational interactions of individual stars emerge the glorious sweeping arms of spiral galaxies. From the chemical interactions of individual ants emerge the extraordinarily complex social behavior of ant colonies. From the electrical interactions of individual neurons in your brain emerge thought and self-awareness. Emergence is nature's most powerful tool for making the universe a complex, patterned, entertaining place to live.

Life itself is arguably the most remarkable of all emergent systems. Many origin-of-life experts adopt the view that life began as an inexorable sequence of emergent events, each of which was an inevitable consequence of interactions among versatile carbon-based molecules. Each emergent episode added layers of chemical and structural complexity to the existing environment. Intensive experiments at laboratories around the world reveal, step-by-step, the essential life-triggering reactions that must occur throughout the cosmos. First came the carbon-containing biomolecules, synthesized in unfathomable abundance on comets and asteroids, in the black near-vacuum of space, on the surface of the young Earth, and deep within our planet's restless crust. Then came the emergence of larger molecular structures—the selection, concentration, and assembly of life's membranes, proteins, and genetic molecules, built in part on a scaffolding of rocks and minerals. Eventually, these biomolecular structures formed self-replicating cycles—chemical systems that copied themselves and competed for a finite and dwindling supply of resources. Ultimately, competition between different self-replicating cycles triggered evolution by natural selection, and life was on its way.

Part I introduces the theory of emergence, which provides a conceptual framework for understanding the immensely complex path from nonlife to the first living cell. Parts II, III, and IV explore experimental and theoretical attempts to understand, step by step, the specific emergent processes: the emergence of diverse biomolecules, the emergence of larger structures composed of many molecules, and finally the emergence of self-replicating collections of molecules. Be warned. Not all of these attempts are success stories: Scientific progress demands time-consuming, often tedious effort; most experiments end in failure; we spend much of our scientific lives making mistakes as fast as we can and desperately trying not to make the same mistake twice.

The narrative focuses on the chemical transition from a prebiotic Earth rich in organic molecules to the so-called RNA World of self-replicating genetic molecules. I do not examine the important subsequent transition from the RNA World to the present world in which life is governed and sustained by DNA and proteins, nor do I examine the rich subject of life's evolution following the appearance of the first cell—both are complex topics requiring their own books.

As in any scientific field, much of the richness of origin-of-life research lies in the details and not the overview. Thus I provide an extensive section of notes and a bibliography that directs readers to primary literature and more detailed discussions of many issues. The notes include comments and corrections from numerous experts who reviewed drafts of this book; in many instances these remarks highlight current uncertainties and controversies among scientists, including disagreements about my interpretations. Nonetheless, this book is not encyclopedic and only begins to address a vast and rapidly expanding literature. I apologize to those scientists whose important studies are not detailed.

Why should an earth scientist, trained in the fields of mineralogy and crystallography, write such a book? By the mid-1990s, my research career at the Carnegie Institution of Washington's Geophysical Laboratory had reached a respectable plateau, achieved through two decades of solid, serviceable work in the specialized field of high-pressure crystal chemistry. With secure federal funding and a steady stream of publications, scientific life was good, at least on an immediate level; but something, I felt, was missing. The essence of science is the unmatched joy in seeking and finding answers to questions about the natural world, yet by the mid-1980s we had grasped the central principles of how crystals compress. The crystallographic questions I asked seemed increasingly narrow, while the answers provided few surprises. I was ready to try something new.

There is another reason why origins research resonated. Since the late 1980s, I've worked with a colleague at George Mason University, the physicist James Trefil, to effect reform in undergraduate science education. Most undergraduate science requirements are discipline-bound—tied to physics, chemistry, or biology. Jim and I believe that this approach is flawed, providing precious little useful knowledge on which students can build and grow as lifelong science enthusiasts. Rather, we find that non-science majors benefit immeasurably from broadly integrated science courses that deal with overarching scientific principles and their applications to daily life.

Origin-of-life research epitomizes why integrated science education is vital. The pursuit of life's origins, like many other fascinating facets of the natural world, draws deeply on several branches of science—geology, biology, chemistry, physics, astronomy. It is thrilling to integrate the full spectrum of scientific ideas into the pursuit of one fundamental question.

So, when the unanticipated opportunity arose, I jumped at the chance to change research directions and began investigating the ancient question of life's chemical origins. My efforts have gained encouragement and focus through years of contact with theoretical biologist Harold Morowitz, another colleague at George Mason University. Harold enticed me into the origins field, and his innovative ideas and persuasive approach have informed every step I've taken. His edict that "the unfolding of life involves many, many emergences" provides an underlying theme of this book.

Our first experiments, commenced with giddy optimism in the spring of 1996, proved laughably naïve. Our little group at Carnegie's Geophysical Laboratory was unlikely to leapfrog a half-century of dedicated and creative origin-of-life research, yet we did enter the arena with fresh eyes and a lack of crippling preconceptions. Slowly, a framework for tackling the problem, and this book, emerged.

The outline for *Genesis: The Scientific Quest for Life's Origin* crystallized at a September 2000 conference in Modena, Italy, that focused on a single daunting question: "What is life?" Opinions among the hundred assembled scientists, philosophers, and theologians differed dramatically, but the most contentious debates occurred within the ranks of scientists. One aging expert on lipid molecules argued that life began with the first semipermeable lipid membrane, the structure that encloses cells. An authority on metabolism countered that life began with the first metabolic cycle, the process by which all cells convert energy and atoms to useful molecules. On the contrary, claimed several molecular biologists who specialized in genetics, the first living entity must have been an RNA-like genetic system that carried and duplicated biological information. One mineralogist even proposed that life began not as an organic entity, but as a self-replicating mineral.

To a relative newcomer in the field, the unresolved debate was reminiscent of the old tale of the blind men and the elephant. Asked to describe the massive beast, each one's perspective varied, based on which anatomical feature was close at hand—the rough ropelike tail, the mighty treelike legs, the twisting snakelike trunk, and so forth. Each blind man's version was wrong, but each possessed an element of the more complex elephantine truth. Perhaps, I thought, the disparate claims of what constitutes life are likewise mere parts of the more complex truth of life's identity and origin.

Throughout the conception and writing of this book I have benefited from the aid, advice, and encouragement of numerous colleagues. In addition to Harold Morowitz, they include most prominently geochemist George Cody and petrologist Hatten S. Yoder Jr. of the Geophysical Laboratory. I have continued to rely on them in countless ways. Hat Yoder's death in August 2003 was a great personal and professional loss.

I am indebted to many distinguished origin-of-life researchers who read the manuscript and provided invaluable comments and corrections, including Louis Allamandola, Gustav Arrhenius, Graham Cairns-Smith, David Deamer, James Ferris, Friedemann Freund, Thomas Gold, Rosalyn Grymes, Bruce Jakosky, Gerald Joyce, Noam Lahav, Antonio Lazcano, James Miller, Harold Morowitz, David Oldroyd, Norman Pace, Nick Platts, David Ross, Bruce Runnegar, Sara Seager, Jack Szostak, Günter Wächtershäuser, Malcolm Walter, and Nick Woolfe.

Among the numerous other scientists who contributed directly or indirectly to the book through conversations, collaborations, and comments on portions of the manuscript are Bruce Alberts, Jeffrey Bada, Connie Bertka, Robert Downs, Glenn Goodfriend, Stephen Jay Gould, Nora Noffke, David Sholl, Henry Teng, and James Trefil.

The establishment of the NASA Astrobiology Institute (NAI), of which our Carnegie Institution research team was a founding member in 1998, provided a wealth of new colleagues and collaborative ventures. I have benefited immeasurably from interactions with Aravind Asthagiri, Nabil Boctor, Alan Boss, Kevin Boyce, Jay Brandes, Hugh Churchill, Rachel Dunham, Janae Eason, Gözen Ertem, Mary Ewell, Tim Filley, Marilyn Fogel, Patrick Griffen, Chris Hadidiacos, Wes Huntress, Andy Knoll, Jake Maule, Ken Nealson, David Olesh, Doug Rumble, James Scott, Sean Solomon, Andrew Steele, and Jan Toporski.

The staff at the Joseph Henry Press tackled this project with dedication, professionalism, and constant good humor. First and foremost, my deep gratitude to executive editor Stephen Mautner, who embraced the concept of this book and has contributed to every stage of its development. Steve played many roles—enthusiastic reader, attentive confidant, generous friend, patient counselor. His influence appears on every page. Jim Gormley supervised the stylish graphics and design. Production editor Heather Schofield managed the project with grace and efficiency. Freelance editor Sara Lippincott provided an invaluable detailed review of the penultimate draft. Marketing director Ann Merchant developed the effective publicity campaign. I am grateful to all of the JHP staff for their help and guidance.

Margaret Hazen contributed to every facet of this book, through countless hours of intense conversation, discerning critiques of innumerable drafts, and unflagging support and encouragement.

Prologue

"Look at this!"

Harold Morowitz beamed as he hustled into my office. "The dielectric constant of water drops to about 20 at a kilobar and 350 degrees. That's like an organic solvent!" He pulled up a chair to show me the data.

It took me a moment to change gears and realize what he was so excited about. Harold and I both teach undergraduate science courses at George Mason University, where we often discuss biology's "Big Questions," including the chemical processes that might have led to life's origin. For more than two decades Harold had puzzled over the role of water, which poses a persistent problem in origin-of-life scenarios. Water is the medium of life, while carbon, by far the most versatile of all the chemical elements, forms the essential backbone of all biomolecules. Yet researchers find that several key chemical steps in assembling life's carbon-based molecules do not work very well in water. So how could life have started on a wet planet?

One intriguing, though untested, possibility is that life's initial chemical reactions proceed more easily at high pressure and temperature. This idea had received a boost in the late 1970s, when Oregon State University oceanographer Jack Corliss descended to the deep, dark ocean floor in the research submersible *Alvin* and observed astonishing ecosystems at undersea volcanic vents. In these hellish zones, without benefit of sunlight, life has found a way to survive crushing pressures of 1,000 atmospheres—a kilobar—and scalding temperatures greater than 100°C. Perhaps, Corliss and co-workers suggested, life first arose at such hostile extremes, and in total darkness.

Morowitz's dense tabulation held a possible clue. Photocopied from a 1970s text, it recorded variations of water's physical and chemical properties with temperature and pressure. Sure enough, at extreme pressure-cooker conditions water appeared to be a remarkably different liquid from the stuff that comes out of the tap. Harold was onto something. "So maybe Jack Corliss is right—maybe it *is* the vents."

The mainstream origin-of-life community, wedded as they were to the tradition of a globe-spanning ocean of "primordial soup" bathed in sunlight, had rejected this speculation out of hand. In the intervening two decades, no one had bothered to try the relevant high-temperature and high-pressure experiments. Yet the so-called hydrothermal-origins hypothesis was too intriguing and too testable to disappear. In the late 1980s, the German chemist and patent attorney Günter Wächtershäuser put flesh on the bones of this idea by proposing a detailed chemical scenario for origin events in a deep hydrothermal zone rich in sulfide minerals.

Now Harold had uncovered data showing how the physical and chemical properties of water might be very different at extreme conditions from those of everyday experience. Perhaps chemical reactions that fail at Earth's ordinary surface conditions could take place at those extremes. There was only one way to find out.

"So, can you do the experiments?" Harold knew that my nearby research base, the Carnegie Institution's Geophysical Laboratory in Washington, D.C., maintained an arsenal of high-temperature and high-pressure apparatus. The laboratory specialized in chemical reactions at extreme conditions, so he hoped that my colleagues and I might be able to tackle the complex carbon chemistry that underpins the origin of life.

I hardly gave it a second thought. "Sure, why not? It's an easy experiment." I had no idea where that hasty promise would lead.

Harold Morowitz is one of the kindest scientists I know. More than a few scientists wear a veneer of kindness—a *pro forma* geniality that masks an intense and often competitive personality. Those of us whose lives are devoted to indulging our curiosity tend to be a distracted, self-absorbed lot. Harold is different. He smiles winningly at the slightest encouragement and speaks with the calmness and quiet passion of a rabbi or father confessor. He loves a good idea, and shares his richly inventive vision of life's origin without hesitation, without the conventional expectation of clever reciprocation. He came to George Mason University in 1988, after a full and productive career on the biology faculty at Yale, where he made his mark studying energy flow in cells. Harold argues persuasively that modern cells carry hints of life's earliest biochemical processes. Such molecular "fossils" persist in all of us in all our cells he claims, and these molecules point to the chemistry of life's origins. [Plate 1]

Although at the time a novice at origins research, I was delighted for an excuse to collaborate with Harold Morowitz. His name provided instant credibility in a field at times tarnished by questionable data, contentious debates, or even outright quackery. Nevertheless, a successful experiment must be planned with care, and Harold had already spent a lot of time thinking about strategy.

"Let's start with pyruvate," he urged. Pyruvate, an energy-rich, 3-carbon molecule, was a natural choice for Harold, who had spent a lifetime studying metabolism, the processes by which cells gather atoms and energy to sustain themselves, grow, and reproduce. Pyruvate is a key ingredient in every cell's metabolism. Most cells gain energy by splitting the 6-carbon sugar, glucose, into two pyruvate molecules, after which the pyruvate is broken into smaller molecules to release further energy. Morowitz explained that some cells also use pyruvate as a building block to construct larger molecules. So, for example, a pyruvate molecule and a molecule of carbon dioxide (one carbon atom bonded to two oxygen atoms) can react to form a 4-carbon molecule called oxaloacetate, which undergoes further reactions in metabolism.

It's simple math: 3 + 1 = 4. But that reaction never works in water at room pressure, at least not without a complex biological catalyst—a molecule that greatly boosts the reaction rate. In the absence of a cell's sophisticated catalytic chemical machinery, pyruvate tends to break down into fragments with only one or two carbon atoms. Perhaps, Harold suggested, higher pressure and temperature would reverse this trend and induce pyruvate plus carbon dioxide to form oxaloacetate. If so—if we could demonstrate that hydrothermal conditions promote such a key metabolic reaction—then hydrothermal vents would become even more of a focus for the origin of metabolism. Experimentalists dream of such opportunities—a simple experiment with a big potential payoff.

We agreed on a range of temperatures from 150°C to 300°C and pressures from 500 to 2,000 atmospheres—conditions relevant to hydrothermal systems at and below Earth's deep-ocean floor. But achieving such conditions is easier said than done.

It was time to pay Hat Yoder a visit.

Every scientific career rests on the support of colleagues who act as teachers, collaborators, and employers. In my career, Hatten S. Yoder Jr. served in all three roles. He is my scientific hero. Burly, handsome, with big, powerful hands, Hat built his high-pressure lab at Carnegie's Geophysical Laboratory from scratch, shortly after returning from action with the Pacific Fleet in World War II. For more than half a century, he maintained this premier facility, working tirelessly in his quest to understand the origins of rocks. [Plate 1] Even at the age of 75, I knew he would jump at the chance to try something new.

"What P and T?" he asked.

My response was somewhat sheepish. Hat's pressure lab was designed to achieve extreme conditions: pressures of 10,000 atmospheres at more than 1,000°C. Running our proposed experiments at a measly 2,000 atmospheres and 250°C would be like using a blast furnace to bake a cake. But an experiment is an experiment, and after only the slightest raising of eyebrows and barely audible "hmmph!" Hat was ready to go.

High-pressure experiments are not for the faint of heart. Even at pressures of only a few atmospheres, hot gas can cause a nasty explosion. Your pressure cooker sustains no more than a pressure of 2 atmospheres, your car's tires less than 3, and both can blow out violently. Hat's home-built device, aptly called a pressure bomb, worked at thousands of atmospheres with a volume about the size—and with the same explosive power—of a stick of TNT. A catastrophic, explosive failure at those pressures could knock out a corner of the lab building. But no such worries attached to the pyruvate project.

Our experimental strategy relied on a classic metal-capsule technique for studying chemical reactions at high temperatures and pressures: Simply seal reactants into tiny cylindrical gold tubes about the size of a large grain of rice. The soft, chemically inert gold crushes down on the reactants, providing an isolated high-pressure, high-temperature environment.

We crafted our capsules from a cylinder of gold a foot long, like a precious soda straw. I cut the cylinder into 1-inch lengths and welded each piece shut at one end. Water and pyruvate, both liquids, loaded easily into the capsules with a syringe—50 milligrams of water and 3 milligrams of pyruvate, just a droplet. It's difficult to weld shut a tube containing a gas, so we adopted an old experimental trick and used a white powdered chemical called oxalic acid dihydrate, which breaks down to water plus carbon dioxide above 100°C.

My first attempts at welding the gold tubes shut were a mess. I weighed and loaded the reactants, crimped shut the tube's open end, and placed the gold capsule into a vise, with the crimped end peeking out above thin steel jaws. A successful weld requires one smooth flick of the wrist with a carbon-arc welder, a graphite rod the size of a pencil that carries an intense electrical current. The gold is supposed to melt and flow, zipping up the capsule in a fraction of a second. But as my welder heated the gold, a portion of the volatile pyruvate boiled away. A sudden burp of smelly gas blew an ugly, gaping hole in the weld, ruining the carefully weighed ratios of reactants. After a lot of trial and error, I learned to weld one capsule end (using a microscope to see what I was doing), while the other end was immersed in ultracold liquid nitrogen, a frigid –196°C bath that froze the volatile reactants. The welder would erupt into blue-white flame and the gold would sputter and melt, sealing the tube. Eventually my batting average rose above 0.500.

Hat inserted three identical gold capsules into a platinum holder inside a foot-long nickel metal cylinder that would serve as an electric furnace. Decades of experience streamlined his routine. After the capsules are inserted, load the cylinder with ceramic filler rod, thermocouple wires, and ceramic end cap, and pack it all with a fine, sandlike powder of white aluminum oxide. Attach that sample assembly to the "head," a fat steel plug that holds in pressure while providing insulated channels for wires that carry electric current for the heater and temperature sensors. Insert the cylinder and head into the massive metal bomb. Seal the bomb with a giant 6-inch-long nut with a 6-sided head. Tighten the nut with a 3-foot-long 20-pound wrench; use a 3-foot-long pipe as an extension for added torque.

Hat banged away at the unwieldy wrench, making a horrendous racket as he tightened the nut just a bit more for safety. "It's always good to make a lot of noise in the lab," he said. "That way the director will know you're working."

We retreated behind a wall of battleship gray, war-salvage naval armor as Hat opened and closed a bewildering sequence of valves to fill the system with pressurized argon, an inert gas. *Ka-chunk, ka-chunk, ka-chunk!* The argon gas compressor pumped the bomb to 2,000 atmospheres in a matter of minutes. Hat set the computerized furnace controls to ramp the temperature up to 250°C, and we were off. Deep inside the steel bomb the gold tubes were being crushed and heated to conditions similar to those found several miles beneath Earth's surface. In such an extreme environment, the pyruvate was sure to do something interesting. We went to lunch.

Two hours later, we were back in the high-pressure lab to quench the run. When the current shut off, the temperature dropped rapidly, cooling to below 100°C in about a minute or so. Hat released the pressure with a *whoosh*, unscrewed the big nut, and pulled out the sample assembly. He dumped the cylinder's contents into a shallow metal tray: end cap, thermocouple, filler rod, platinum holder, lots of fine white sand, and three fat, shiny gold capsules spilled out. Success! The capsules had held! We itched to know what was inside. Two capsules went into the freezer, while I took the third upstairs for analysis.

There is nothing sophisticated about opening a gold capsule; you simply snip open one end and pour out the contents. First I washed the outside of the capsule in organic solvents to avoid any contamination by machine oil or fingerprints. Then I froze it in liquid nitrogen, so that the contents would not leak out during the snip. I positioned the capsule over a glass vial to catch the gold and its contents. Just a little snip . . . usually does the trick. . . . *Kapow!* The gold weld blasted off into some remote corner of the lab, propelled by the sudden release of what must have been several atmospheres of internal gas pressure. A bit shaken, I dropped the capsule into the bottom of the vial, where it lay dormant for a few seconds. But then it began to hiss and foam as a yellow-brown oily substance frothed out, coating the gold and the glass. A pungent odor not unlike Jack Daniels permeated the lab.

The pyruvate had clearly reacted, but it did not look anything like colorless, odorless oxaloacetate. What had we made?

Time to consult George Cody [Plate 1], a recent arrival at the lab and an organic geochemist trained to analyze messy, oily stuff. George is enthusiastic, loquacious, and—luckily for us—he can't seem to say no. He is also an expert in the chemistry of coal; he tends to snow visitors to his office with a blizzard of arcane chemical names and reactions. He thinks out loud and scribbles diagrams of molecules and reactions on any available surface, including the protective windows of his lab's chemical hoods. When I showed him the smelly goo, he knew just what to do.

"GCMS," he said, "We probably don't need CI." I nodded as if I understood what he was talking about. "Let's use BF_3 propanol as the derivatizing agent. The Supelco column should work fine."

He had proposed that we analyze our suite of products by passing them, together with a chemically inert gas, through a long, thin tube filled with specially prepared organic molecules. This technique, gas chromatography (the "GC" of GCMS), separates different molecules according to how fast they move through the column. In general, smaller, less reactive molecules move faster than bigger, "sticky" molecules. The gas chromatograph sorts a collection of different molecules into separate little pulses, typically over a period of 30 or 40 minutes.

Then comes the mass spectrometer (the "MS" of GCMS), which measures the relative masses of molecules and their fragments. George's mass spectrometer blasts molecules into lots of smaller pieces of distinctive weights, so each pulse from the GC can be analyzed separately as a suite of characteristic mass fragments, providing a kind of fingerprint of the product molecule.

It took a couple of hours of chemical processing to prepare the concentrated liquid sample for analysis. George filled a syringe with the pale yellow liquid and injected a tiny drop into the GCMS with a practiced, swift motion. We sat back to watch as a spectrum gradually appeared on the computer monitor. The first peak showed up at 10.79 minutes—a small molecule with probably only two or three carbon atoms. Then another peak at 11.71 minutes, and another at 11.96. Faster and faster peaks appeared, piling in on top of each other, every spike representing additional molecular products. By the 20-minute mark, a broad hump decorated with hundreds of sharp spikes was emerging.

"Humpane," George muttered in disgust. Pyruvate had reacted in our capsules, to be sure. But instead of the simple 3 + 1 = 4 reaction that Morowitz had proposed, we had produced an explosion of molecules—tens of thousands of different kinds of molecules. Not a trace of oxaloacetate was to be found, but a bewildering array of other molecular species had emerged. It might take a lifetime to decipher the contents of just one such molecular suite.

One conclusion was obvious. Some very dynamic organic reactions proceed rapidly at hydrothermal conditions. In one sense, Morowitz's hypothesis had failed: Pyruvate doesn't react with carbon dioxide to form oxaloacetate under those conditions. But we had caught our first glimpse of a robust, emergent carbon chemistry in a hydrothermal environment. This was chemistry worth exploring.

Where to begin? We were faced with choosing from among thousands of simple carbon-based molecules over a wide range of pressure, temperature, and other experimental variables—work to devour a hundred scientific lifetimes.

What were we getting ourselves into?

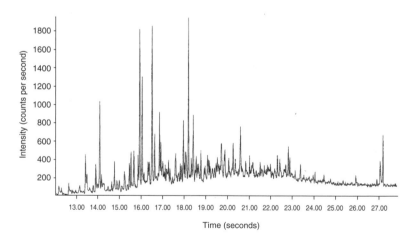

A diverse suite of molecules emerges when pyruvate is subjected to high temperature and pressure. These products appear as numerous sharp peaks superimposed on a broad "humpane" feature on a gas chromatogram.

Part I

Emergence and the Origin of Life

All origin-of-life researchers face the baffling question of how the biochemical complexity of modern living cells emerged from a barren, primordial geochemical world. The only feasible approach is to reduce biological complexity to a comprehensible sequence of chemistry experiments that can be tackled in the human dimensions of space and time—a lab bench in a few weeks or months. George Cody, Hat Yoder, and I were eager to continue our hydrothermal experiments, but what should come next? We knew that the simplest living cell is intricate beyond imagining, because every cell relies on the interplay of millions of molecules engaged in hundreds of interdependent chemical reactions. Human brains seem ill suited to grasp such multi-dimensional complexity.

Scientists have devised countless sophisticated chemical protocols, and laboratories are overflowing with fancy analytical apparatus. Chemists have learned to synthesize an astonishing array of paints, glues, cosmetics, drugs, and a host of other useful products. Yet when confronted with the question of life's ancient origin, it's easy to become mired in the scientific equivalent of writer's block. How does one begin to tackle the chemical complexity of life?

One approach to understanding life's origin lies in reducing the living cell to its simpler chemical components, the small carbon-based molecules and the structures they form. We can begin by studying relatively simple systems and then work our way up to systems of greater complexity. In such an endeavor, the fascinating new science of emergence points to a promising research strategy.

1

The Missing Law

It is unlikely that a topic as complicated as emergence will
submit meekly to a concise definition, and I have no such
definition.
John Holland, *Emergence: From Chaos to Order*, 1998

Hot coffee cools. Clean clothes get dirty. Colors fade. People age
and die. No one can escape the laws of thermodynamics.

Two great laws, both codified in the nineteenth century by a small
army of scientists and engineers, describe the behavior of energy. The
first law of thermodynamics establishes the conservation of energy.
Energy, which is a measure of a system's ability to do work, comes in
many different forms: heat, light, kinetic energy, gravitational potential, and so forth. Energy can change from any one form to another
over and over again, but the total amount of energy does not change.
That's the first law's good news.

The bad news is that nature places severe limitations on how we
can use energy. The second law of thermodynamics states that heat
energy, for example, always flows from warmer to cooler regions, never
the other way, so the concentrated heat of a campfire or your car's
engine gradually radiates away. That dissipated heat energy still exists,
but you can't use it to do anything useful. By the same token, all natural systems tend spontaneously to become messier—they increase in
disorder, or "entropy." So any collection of atoms—be it your shiny
new shoes or your supple young body—gradually deteriorates. The
second law of thermodynamics is more than a little depressing.

But look around you. You'll find buildings, books, automobiles,
bees—all of them exquisitely ordered systems. Despite the second law's

dictum that entropy increases, disorder is not the only end point in the universe. Observations of such everyday phenomena as sand dunes, seashells, and slime mold reveal that the two laws of thermodynamics may not tell the entire story. Indeed, some scientists go so far as to claim that a fundamental law of nature, the law describing the emergence of complex ordered systems (including every living cell), is missing from our textbooks.

THE LAWS OF NATURE

The discovery of a dozen or so natural laws represents the crowning scientific achievement of the past four centuries. Newton's laws of motion, the law of gravity, the laws of thermodynamics, and Maxwell's equations for electromagnetism collectively quantify the behavior of matter, energy, forces, and motions in almost every human experience. The power of these laws lies in their universality. Each law can be expressed as an equation that applies to an infinite number of events, from the interactions of atoms to the formation of galaxies. Armed with these laws, scientists and engineers confidently analyze almost any physical system, from steam engines to stars.

So sweeping and inclusive are these natural laws that some scholars of the late nineteenth century suggested that the entire theoretical framework of science had been deduced. All that remained to be discovered were relatively minor details, like filling in the few remaining gaps in a stamp collection. Though this turned out not to be the case—modern physics research has revealed new phenomena at the relativistic scales of the very small, the very fast, and the very massive—the classic laws do indeed still hold sway in our everyday lives.

Yet in spite of centuries of labor by many thousands of scientists, we do not fully understand one of nature's most transforming phenomena—the emergence of complexity. Systems *as a whole* do tend to become more disordered with time, but at the *local scale* of a cell, an ant colony, or your conscious brain, remarkable complexity emerges. In the 1970s, the Russian-born chemist Ilya Prigogine recognized that these so-called complex emergent systems arise when energy flows through a collection of many interacting particles. The arms of spiral galaxies, the rings of Saturn, hurricanes, rainbows, sand dunes, life, consciousness, cities, and symphonies all are ordered structures that emerge when many interacting particles, or "agents"—be they molecules, stars, cells, or people—are subjected to a flow of energy. In the jargon of thermodynamics, the formation of patterns in these systems helps to speed up the dissipation of energy as mandated by the second law. Scientists and nonscientists alike tend to value the surprising order and novelty of such emergent systems.

The recognition and description of such emergent systems provides a foundation for origin-of-life research, for life is the quintessential emergent phenomenon. From lifeless molecules emerged the first living cell. If we can understand the principles governing such systems, we may be able to apply those insights to our experimental programs.

DESCRIBING EMERGENT SYSTEMS

If you want to enunciate a law that characterizes emergent systems, then the first step is to examine everyday examples. You can observe emergent behavior in countless systems all around us, including the interactions of atoms, or of automobiles, or of ants. This universal tendency for systems to display increased order when lots of objects interact, while fully consistent with the first and second laws of thermodynamics, is not addressed explicitly in either of those laws. We have yet to discover if all emergent systems possess a unifying mathematical behavior, though our present ignorance should not seem too unsettling. It took more than a half-century for each of the first two laws of thermodynamics—describing the behavior of energy and entropy, respectively—to develop from qualitative ideas into quantitative laws. I suspect that a mathematical formulation of emergence will be discovered much sooner than that, perhaps within the next decade or two.

Scientists have already identified key aspects of the problem. Many familiar natural systems lie close to equilibrium—that is, they are stable and unchanging—and thus they do not display emergent behavior. Water gradually cooled to below the freezing point equilibrates to become a clear chunk of ice. Water gradually heated above the boiling point similarly equilibrates by converting to steam. For centuries, scientists have documented such equilibrium processes in countless carefully controlled scientific studies.

Away from equilibrium, dramatically different behavior occurs. Rapidly boiling water, for example, displays complex, turbulent convection. Water flowing downhill in the gravitational gradient of a river valley interacts with sediments to produce the emergent landform patterns of braided streams, meandering rivers, sandbars, and deltas. These patterns arise as energetic water moves.

Emergent systems seem to share this common characteristic: They arise away from equilibrium when energy flows through a collection of many interacting particles. Such systems of agents tend spontaneously to become more ordered and to display new, often surprising behaviors. And as patterns arise, energy is dissipated more efficiently, in accord with the second law of thermodynamics. Ultimately, the resulting behavior appears to be much more than the sum of the parts.

Emergent patterns in water and sand may seem a far cry from living organisms, but for scientists studying life's origins there's a big payoff in understanding such simple systems: Of all known emergent phenomena, none is more dramatic than life, so studies of simpler emergence can provide a conceptual basis, a jumping-off point, for origin-of-life research.

QUANTIFYING THE COMPLEXITY OF EMERGENT SYSTEMS

Even though emergent systems surround us, a rigorous definition (much less a precise mathematical formulation) remains elusive. If we are to discover a natural law that describes the behavior of emergent systems, then we must first identify the essential properties of such systems. But what characteristics distinguish emergent systems from other less interesting collections of interacting objects?

All emergent systems display the rather subjective characteristic of "complexity"—a property that thus far lacks a precise quantitative definition. In a colloquial sense, a complex system has an intricate or patterned structure, as in a complex piece of machinery or a Bach fugue. "Complexity" may also refer to information content: An advanced textbook contains more detailed information, and is thus more complex, than an elementary one. In this sense, the interactions of ants in an ant colony or neurons in the human brain are vastly more complex than the behavior of a pile of sand or a box of Cheerios.

Such complexity is the hallmark of every emergent system. What scientists hope to find, therefore, is an equation that relates the properties of a system on the one hand (its temperature or pressure, for example, expressed in numbers), to the resultant complexity of the system (also expressed as a number) on the other. Such an equation would in fact be the missing "law of emergence." But before that is possible we need an unambiguous, quantitative definition of the complexity of a physical system. How to proceed?

A small band of scientists, many of them associated with the Santa Fe Institute in New Mexico, have thought long and hard about complex systems and ways to model them mathematically. But their efforts yield surprisingly diverse (some would say divergent) views on how to approach the subject.

John Holland, an ace at computer algorithms and a revered founder of the field of emergence, models emergent systems as computer programs with a fixed set of operating instructions. He suspects that any emergent phenomenon, including sand ripples, ant colonies, the conscious brain, and more, can be reduced to a set of selection rules. Holland and his followers have made great strides in mimicking natural phenomena with a few lines of computer code. Indeed, for Holland and his followers the complexity of a system is closely related to the minimum number of lines of computer code required to mimic that system's behavior.

A delightful example of this approach is BOIDS, a simple program written by California programmer Craig Reynolds that duplicates the movements of flocking birds, schooling fish, swarming insects, and other collective animal behaviors with astonishing accuracy. (To check it out on the Internet, just Google "BOIDS.") Lest you think that this effort is idle play, remember that computer programmers of video games and Hollywood special effects have made a bundle on this type of simulated emergent behavior. Think of BOIDS the next time you watch dinosaur herds on the run in *Jurassic Park*, swarming locusts in *The Mummy*, or schools of fish in *Finding Nemo*.

Physicist Stephen Wolfram, a mathematical prodigy who made millions in his twenties from the elegant, indispensable computer package *Mathematica*, provides a complementary vision of emergent complexity from simple rules. Like Holland, Wolfram was captivated by the power of simple instructions to generate complex visual patterns. Sensing a new paradigm for the description and analysis of the natural world, he has spent the past 20 years developing what he calls "a new kind of science" (NKS for short). A mammoth tome by that title published in 2002 and an elaborate Web site (www.wolframscience.com) illustrate some of the stunning ways whereby geometric complexity may arise from simple rules. Perhaps, Wolfram argues, the complex evolution of the physical universe and all it contains can be modeled as a set of sequential instructions.

Many other ways to view complex systems have been proposed. The late Danish physicist Per Bak described complex systems in terms of a mathematical characteristic called "self-criticality." These systems evolve by repeatedly achieving a critical point at which they falter and regroup, like a growing pile of sand that avalanches over and over again as new grains are added. Santa Fe theorist Stuart Kauffman proposes another tack, focusing on the emergence of chemical complexity via competitive "autocatalytic networks," by which collections of chemical compounds catalyze their own formation. And Nobel laureate Murray Gell-Mann, who also works at the Santa Fe Institute, has recently introduced a new parameter he calls "nonextensive entropy"— a measure of the intrinsic complexity of a system—as a path to understanding complex systems.

All these approaches and more inform the search for a law of emergence; all provide a glimpse of the answer. Yet each seems too abstract to apply to benchtop chemical experiments on the origin of life. An experimentalist needs to decide on the nitty-gritty details: What should be the starting chemicals at what concentrations; how acidic or basic the solution; what run temperatures, pressures, and times? Is there any way that the ideas of emergence can help?

A classic scientific approach to discovering general principles and laws is to examine the behavior of specific systems. The study of simple systems that display emergent behavior may well point to physical factors that lead to patterning in much more complex systems, including life. We can hope that observations of specific systems will eventually point to more general rules.

PATTERNS IN THE SAND

You don't need a laboratory to observe emergent phenomena. In fact, you can't go on a hike without seeing dozens of examples of emergence in action. Among my favorite emergent phenomena are interactions of water and sand, which provide a convenient and comprehensible example of structures arising from the energetic interactions of lots of agents (not to mention a great excuse to spend the day at the shore). When moving water (or wind, for that matter) flows across a flat layer of sand, new patterns arise. Periodic sand ripples appear, as sand grains are sorted by size, shape, and density. The system thus becomes more orderly and patterned as energy—the flow of wind or water—dissipates.

My favorite emergent sandy system lies at the base of the fossil-rich hundred-foot-tall cliffs that border the Chesapeake Bay's western shore in Calvert County, Maryland. Fifteen-million-year-old whale bones, razor-sharp sharks' teeth, branching bleached corals, and robust fist-sized clamshells abound in the wash zone, where waves constantly wear away the soft sediments. Walks along those majestic formations often lead to thoughts about the factors that contribute to complexity.

At times of unusually low tide, especially near a new moon in the cold clear winter months, receding waters expose a gently sloping pavement of ancient sediments below the base of the cliff—a formation called blue marl. Treacherously slippery when wet, this firm flat surface commonly accumulates a thin layer of sand—particles that display emergent patterns when subjected to the wash of shallow water. Over the years, I've noticed four distinct factors that contribute to the emergence of complex sand patterning.

Factor 1: The Concentration of Agents

The first obvious factor in achieving a patterned, complex system is simply the density of sand grains—that is, the number of interacting particles per square centimeter of the blue marl's surface. It's easy to estimate this number by collecting almost every grain of sand from an area 10 centimeters square, about the size of a small paper napkin. I collect the sand in a plastic bag or bottle, take it back to the lab, dry it, and weigh the sample. Using a microscope, I count out 100 grains from the sample and then weigh that batch. As it turns out, the total number of grains per square centimeter is approximately equal to the total weight of sand from the 100 square-centimeter (10×10) area divided by the weight of 100 sand grains.

I find that with fewer than about 100 sand grains per square centimeter, the dusting of particles is too sparse for any noticeable patterns to emerge. Given the minute size of the average sand grain, typically less than half a millimeter in diameter, 100 grains per square centimeter provides a sparse coverage over less than 10 percent of the smooth

A

B

C

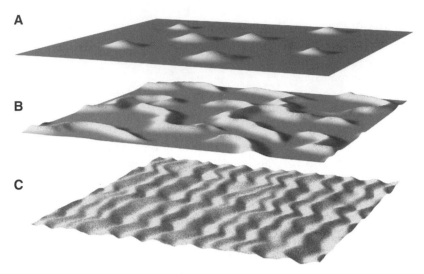

Patterns in sand grains emerge as the concentration of grains increases. At about a thousand grains per square centimeter (A) small, black-topped piles are observed; at a few thousand grains per square centimeter (B) discontinuous bands arise; and above 10,000 grains per square centimeter (C) continuous ripples cover the surface.

blue marl surface. Increase the sand concentration to about 1,000 grains per square centimeter, however, and an intriguing pattern of regularly spaced sand piles, each a centimeter or two across, appears on the hard blue surface. What's more, a small circle of darker sand grains typically crowns each little tan pile. Evidently a minimum concentration of several hundred grains per square centimeter is required to initiate patterning in sand.

Increase the sand concentration slightly to a few thousand grains per square centimeter and you get discontinuous short bands of sand at right angles to the gentle back-and-forth wave motion of the shallow water. As with the mini-sandpiles, each tan band is topped by a line of darker grains. And as sand concentration exceeds 10,000 grains per square centimeter, continuous, evenly spaced, black-capped ripples form across the hard pavement. I've seen this classic rippled surface cover hundreds of square meters of shallow water in patterns so hypnotically regular that I hesitated to disturb the symmetry by walking on it.

And that's it. Higher concentrations of sand simply provide a deeper base for the regular ripples. Buried sand grains don't participate in the process so no new structures arise beyond the elegant, wave-like, periodic forms on the surface.

This systematic behavior suggests that the concentration of interacting agents plays a fundamental role in the emergent complexity of a· system. Below a critical threshold, no patterns are seen. As particle concentrations increase, so too does complexity, but only to a point. Above a critical saturation of agents, we find no new behaviors.

Similar observations have been made about other emergent systems. One ant species—*Eciton burchelli*, the army ant—stays close to home as long as the colony consists of fewer than about 80,000 individuals. Exceed that number of army ants, however, and the colony exhibits new emergent behavior; like a bursting dam, the ants pour out in a massive "swarm raid" to attack adjacent colonies. At higher populations, half of the ants may spontaneously leave to form a new colony. Studies of termite colonies also reveal that the construction of pillar-type mounds requires a critical density of individuals.

At a much greater scale, spiral galaxies require a minimum number of about 100 million stars to trigger development of the familiar spiral arm structure. According to theoretical models of astrophysicists, the majestic arms form as a result of gravitational instabilities caused in part by a large central mass of stars.

Human consciousness and self-awareness also emerge from the interactions of trillions of neurons. Sadly, as those of us who watch friends and relatives afflicted with Alzheimer's disease must observe, when a critical number of cells and their connections are destroyed, self-awareness fades away.

These findings suggest that the emergence of life might have depended on achieving some minimal concentration of biomolecules, the essential agents of cellular life. Too few molecules, no matter how friendly the environment, and life could not arise. That's a useful idea to bear in mind when designing origin-of-life experiments.

Factor 2: The Interconnectivity of Agents

Sand grains influence each other by direct contact, the simplest local way to interact. A rounded grain at the surface of a sandpile typically touches about a half-dozen adjacent grains. The balance between these stabilizing contacts and gravity on the one hand, and the restless, disruptive flow of water on the other, leads to a controlled shuffling of grains and ultimately to the rippled patterning of sand. By contrast, ants in an ant colony interact over much greater distances, by marking the ground with a variety of pheromones, which are chemical signals that point other ants to food, alert them to danger, and provide other vital information. In this way, any given ant has the potential to interact with thousands of colony mates in varied ways. These differences in interconnectedness provide part of the reason why ant colonies are more complex than water-shaped sandpiles.

The conscious brain, the most complex system we know, is also the most complexly interconnected. Each of the trillions of neurons in your brain interacts with hundreds of other nearby cells through a branching network of dendrites. Electrical signals between any two neurons, furthermore, may be stronger or weaker, like the current controlled by the dimmer switch on your lamp. Interconnections of the brain are vastly more intricate than those of sand or ants.

These observations of emergent systems suggest that life's origin must have relied on a wide repertoire of chemical interactions. Experiments that optimize the number and type of molecular contacts might thus be more likely to display emergent behaviors of interest.

Factor 3: Energy Flow Through the System

Regardless of how many sand grains or ants or neurons are present, no pattern can emerge without a flow of energy through the system. Sand grains will not start hopping without a certain minimum water-wave speed (typically about $1/_2$ to 1 meter per second along the shores of the Chesapeake Bay). More energetic waves with greater speed and amplitude move grains more easily and generate sand patterns more quickly, though these patterns do not appear to differ fundamentally in their shapes.

But every complex patterned system has a limit to the magnitude of energy flow it can tolerate. During energetic storms, crashing waves obliterate sand ripples and other local sedimentary features. Black and tan sand grains become jumbled and all signs of emergent patterning disappear.

The human brain exhibits strikingly similar behavior in terms of energy flow. During normal waking hours, the brain maintains a moderate level of electrical impulses—the normal healthy flow of energy through the neural system. Deepest sleep corresponds to a sharp drop in electrical activity as we slip from consciousness, whereas the excessive electrical intensity of an epileptic seizure thwarts conscious action by scrambling the usual patterned electrical flow.

The emergence of complex patterns evidently requires energy flow within rather restrictive limits: Too little flow and nothing happens; too much flow and the system is randomized—entropy triumphs. This conclusion is important for the experimental study of life's chemical origins. A reliable source of energy is essential, to be sure, but lightning, ultraviolet radiation, and other intense forms of ionizing energy can blast molecules apart and may be too extreme to jump-start life. We must look for gentler chemical energy sources, like the steady, reliable chemical potential energy stored in a flashlight battery, to sustain the metabolism of primitive life.

Factor 4: Cycling of Energy Flow

Many natural systems are subject to cycles of energy: day and night, summer and winter, high tide and low tide. Such cycles may play a fundamental role in the evolution of emergent systems, though it's often difficult to document the effects of these subtle cycles in nature.

Laboratory wave tanks, though considerably less scenic than the Chesapeake Bay in January, facilitate the study of sand-ripple formation under controlled conditions. Recent research on natural patterned systems reveals that cycling of energy flow through a system is a fascinating and previously unrecognized fourth factor in generating complex sand patterns. In 2001, physicist Jonas Lundbek Hansen at the Niels Bohr Institute and his Danish colleagues announced this surprising wrinkle in the mechanics of ripple formation. Most previous experiments had involved fixed wave amplitudes (that is, wave height) and frequencies (how many waves pass a given point in a second). Such studies typically generate perfectly spaced, straight ripples. Instead, Hansen and his colleagues wondered what might happen if they cycled these variables. Over periods of several minutes, they increased and then decreased the amplitude or the frequency of their water waves. The results were breathtaking. Rather than simple parallel sand ripples, they produced elegant intertwined and branching sand structures. These new patterns appear remarkably similar to sand features that commonly arise along the Chesapeake Bay when the water is only a few inches deep—conditions that apparently favor periodic fluctuations in wave amplitude.

Alert to the potential power of energy cycling, PhD student Mark Kessler and Professor Brad Werner of the University of California, San Diego, recently analyzed amazing stone circles and other so-called "patterned grounds" in Alaskan Arctic terrain that is subject to cyclical freezing and thawing. With each thaw, rounded boulders shift slightly, interacting with one another over many years to produce remarkable fields covered by natural circles of stone. [Plate 2]

The role of cycling in the emergence of patterns represents a frontier area of study that is keenly watched by some origin-of-life investigators. After all, the primitive Earth was subject to many cycles—day/night, high tide/low tide, wet/dry, and more. Perhaps such cycles, which can be duplicated in a controlled experimental environment, contributed to the emergence of life itself.

FORMULATING EMERGENCE

So what might a mathematical law of emergence look like? My guess is that the expression will take the form of a mathematical inequality, something like this:

$$C \leq f[n, i, \nabla E(t)]$$

That's a short-hand way of saying that the emergent complexity of a system, denoted by the letter C (for "complexity"), is a number less than or equal to some value that is a mathematical function (f) of the concentration of interacting particles (n), the degree of those particles' interconnectivity (i), the time-varying energy flow through the system [$\nabla E(t)$], and perhaps other variables as well.

At least two daunting impediments thwart the completion of this potentially simple formulation. First, as previously noted, we lack a precise definition of complexity. It's impossible to quantify something when you don't really know what that something is. And second, we are woefully ignorant of the exact mathematical relationships between complexity and the three possible key factors: the concentration of interacting agents, the interconnectivity of those agents, and the cyclical energy flow. Simple systems yield tantalizing clues, but we are still a long way from any definitive formula.

This quest to characterize emergent phenomena, though initially couched in mathematical abstraction, is not ultimately an abstract exercise. Emergent systems frame every aspect of our experience. Our environment, our bodies, our minds, the patterns of our lives and our culture—all display emergent complexity. A comprehensive theory of emergence will foster applications to myriad problems in everyday technology: long-range weather prediction, computer network design, traffic control, the stabilization of ecosystems, the control of epidemics, perhaps even the prevention of war. Armed with such a law, we will acquire a deeper understanding of any system of many interacting agents—indeed, even of the origin of life itself.

2

What Is Life?

I know it when I see it.
                    Justice Potter Stewart, 1964

A recent origin-of-life text features an appendix with scientific definitions of life written by 48 different authorities. The entry contributed by the distinguished evolutionary biologist John Maynard Smith describes life as "any population of entities which has the properties of multiplication, heredity and variation." Alternatively, information theorist Stuart Kauffman claims that "life is an expected, collectively self-organized property of catalytic polymers." Other equally renowned experts propose that "Life is the ability to communicate," "Life is a flow of energy, matter and information," "Life is a self-sustained chemical system capable of undergoing Darwinian evolution." The definitions go on and on. Remarkably, no two definitions are the same.

This lack of agreement represents an obvious problem for those who search for signs of living organisms on other worlds, as well as for origin-of-life researchers. It is difficult to be sure that you've discovered life—or deduced the process of life's origin, for that matter—when you can't define what it is. In spite of generations of work by hundreds of thousands of biologists, in spite of countless studies of living organisms at every scale from molecules to continents, we still have no widely accepted definition.

This frustrating lack is not particularly surprising. For one thing, the question "What is life?" is asked in different contexts by different professions. Theologians hotly debate it in relation to the beginning of human life. Does life start at the moment of conception, when the fetal brain first responds, or when the unborn heart first beats? In some theologies, life commences not with a physical process, but at the unknowable supposed instant of ensoulment. At the other end of the human journey, doctors and lawyers require a definition of life in order to deal ethically with patients who are brain dead or otherwise terminally unresponsive.

In contrast to these ethically complex and emotionally charged issues are the more abstract scientific efforts to define life. Biologists rely on straightforward genetic analysis—tests for DNA or diagnostic proteins—to identify the presence of life-forms on Earth today. But a more general definition that distinguishes all imaginable living objects from the myriad nonliving ones remains elusive. We know relatively little about the diversity of cellular life on Earth, not to mention the vast range of plausible noncellular life-forms that might await discovery elsewhere in the universe. Endorsing a sweeping definition of life based on such scanty knowledge is akin to defining "music" after listening to a single recording of Bach's solo cello suites over and over again. The suites are a sublime example of music, but hardly sufficient to characterize the entire genre.

"TOP-DOWN" VERSUS "BOTTOM-UP"

Scientists crave an unambiguous definition of life, and they adopt two complementary approaches in their efforts to distinguish that which is alive from that which is not. Many scientists adopt the "top-down" approach. They scrutinize all manner of unambiguous living and fossil organisms to identify the most primitive entities that are, or were, alive. For origin-of-life researchers, primitive microbes and ancient microfossils have the potential to provide relevant clues about life's early chemistry. This strategy is limited, however, because all known lifeforms, whether living or fossil, are based on biochemically sophisticated cells containing DNA and proteins. Any definition of life based on top-down research is correspondingly limited.

By contrast, a small army of investigators pursues the so-called "bottom-up" approach. They devise laboratory experiments to mimic the emergent chemistry of ancient Earth environments. Eventually, the bottom-up goal is to create a living chemical system in the laboratory from scratch—an effort that might clarify the transition from nonlife to life. Such research leads to an amusing range of passionate opinions regarding what is alive, because each scientist tends to define life in terms of his or her own chosen specialty. One group will focus on the origin of cell membranes; to them, life began when the first encapsulating membrane appeared. Another team studies the emergence of metabolic cycles, so naturally for them the origin of life coincided with the origin of metabolism. Still other groups investigate primordial RNA (DNA's presumed precursor genetic material), viruses, or even artificial intelligence, and each group hawks its own definition of life's first appearance.

Into this mix, philosophers and theologians inject a more abstract view and speculate on the full range of phenomena that might be said to be alive—robotic life, computer life, even a self-aware Internet. Such debates can at times sound like a science fiction convention, but defining life is no idle exercise. The scientific community, with the full support of NASA and other governmental agencies, holds regular meetings to debate the question. After all, one of NASA's prime missions is to look for life on other worlds, so a clear definition is essential for planning future missions.

It's amazing how the "What is life?" question sparks arguments and fosters hard-line positions. Scientists excel at many things, but compromise is not always one of them. Nevertheless, Gerald Joyce of The Scripps Research Institute, serving on a NASA Exobiology panel, proposed a widely cited "working definition" for life in the context of space exploration. "Life is a self-sustained chemical system capable of undergoing Darwinian evolution," he suggested.

According to this opinion, life combines three distinctive characteristics. First, any form of life must be a chemical system. Computer programs, robots activated with microchips, or other electronic entities are not alive according to this definition. Life also grows and sustains itself by gathering energy and atoms from its surroundings—the essence of metabolism. Finally, living entities must display variation. Natural selection of the more fit individuals will inevitably lead to evolution and the emergence of more complex entities. This NASA-inspired definition is probably as general, useful, and concise as any we are likely to come up with—at least until we discover more about what is actually out there.

Even armed with this functional definition, it's difficult to know what Earth's very first life-form was like. Our planet's earliest life may have been vastly different from anything we know today. Many experts suspect that the first living entity was not a single isolated cell as we know it, for even the simplest cell incorporates astonishing chemical complexity. That first life-form probably did not use DNA, given the exceedingly intricate genetic mechanism of life on Earth today. It may not even have used proteins, the chemical workhorses of cellular life.

Experts in different fields propose different ideas regarding Earth's first life-form. As a geologist, trained in the ways of rocks, my favorite theory is that the very first entity to fit NASA's trial definition may have been an extremely thin molecular coating on rock surfaces. Such "flat life" would have spread across mineral grains in a layer only a few billionths of a meter thick, exploiting energy-rich mineral surfaces while slowly spreading like a lichen from rock to rock.

Whatever the first life-form looked like, it must have arisen from chemical reactions of ocean, atmosphere, and rocks. Yet the overarching problem with studying life's origin is that even the simplest known life-form is vastly more complex than any nonliving components that might have contributed to it. How does such astonishing, intricate complexity arise from lifeless raw materials? Emergence can help.

ORIGINS AND EMERGENCE

French anthropologist Claude Lévi-Strauss, who investigated the mythologies of many cultures, identified a deep-seated human tendency to reduce complex situations to oversimplified dichotomies: friend and enemy, heaven and hell, good and evil. The history of science reveals that scientists are in no way immune to this mindset. In the eighteenth century, the neptunists, who favored a watery origin for rocks, fought with the plutonists, who favored heat as the causative agent. Both, it turns out, were right. A similar contentious and ultimately misleading dichotomy raged between the eighteenth-century catastrophists and uniformitarians, the former espousing a brief and cataclysmic geological history for Earth and the latter holding that geological processes are gradual and ongoing. Once-doctrinal distinctions between plants and animals or between single-celled and multicellular organisms have become similarly blurred.

Attempts to formulate an absolute definition that distinguishes between life and nonlife represents a similar false dichotomy. Here's why. The first cell did not just appear, fully formed with all its chemical sophistication and genetic machinery. Rather, life must have arisen through a sequence of emergent events—diverse processes of organic synthesis, followed by molecular selection, concentration, encapsulation, and organization into diverse molecular structures. The emergence of self-replicating molecules of increasing complexity and mutability led to molecular evolution through the process of natural selection, driven by competition for limited raw materials. That sequential process is an organizing theme of this book.

What appears to us as a yawning divide between nonlife and life obscures the fact that the chemical evolution of life occurred in this stepwise sequence of successively more complex stages of emergence. When modern cells emerged, they quickly consumed virtually all traces of the earlier stages of chemical evolution. "Protolife" became a rich source of food, wiped clean by the consuming cellular life, like a clever murderer leaving the scene of the crime.

Our challenge, then, is to play detective—to establish a progressive hierarchy of emergent steps leading from a prebiotic Earth enriched in organic molecules, to functional clusters of molecules perhaps arranged on a mineral surface, to self-replicating molecular systems that copy themselves using resources in their immediate environment, to encapsulation in membranes—that is, to cellular life. (Recall the words of Harold Morowitz: "The unfolding of life involves many, many emergences.") The nature and sequence of these steps may vary in different environments, and we may never know the exact sequence (or sequences) that occurred on the early Earth. Yet many of us suspect that the inexorable direction of the chemical path is similar on any habitable planet or moon.

Such a stepwise scenario informs attempts to define life. To define the exact point at which such a system of gradually increasing complexity becomes "alive" is intrinsically arbitrary. Where you, or I, or anyone else chooses to draw such a line is more a question of perceived value than of science. Do you value the intrinsic isolation of each living thing? Then for you, life's origin may correspond to the entrapment of chemicals by a flexible cell-like membrane. Or is reproduction—the extraordinary ability of one creature to become two and more—your thing. Then self-replication becomes the demarcation point. Many scientists value information as the key and argue that life began with a genetic mechanism that passed information from one generation to the next.

"What is life?" is fundamentally a semantic question, a subjective matter of taxonomy. Nature holds a rich variety of complex, emergent chemical systems, and scientists increasingly are learning to craft such systems in the laboratory. No matter how curious or novel their behavior, none of these systems comes with an unambiguous label: "life" or "nonlife."

To be sure, labels are important and scientists convene earnest conferences and appoint august committees to decide on taxonomic issues. Valid taxonomy is vital for effective communication and provides a foundation for any scientific pursuit. The problem facing us today, however, is that valid taxonomies rely on a minimum level of understanding. Early attempts at classifying animals purely by color, shape, or other superficial features ultimately failed. Similarly, the classification of chemical elements by their physical state—solid, liquid, or gas— was unhelpful in developing a predictive chemical theory.

Recently, the philosopher Carol Cleland of the University of Colorado and the planetary scientist Christopher Chyba of the SETI (Search for Extraterrestrial Intelligence) Institute compared current attempts to define life with similar eighteenth-century efforts to characterize water. Before the discovery of molecules and atomic theory, water could be characterized only by a series of non-unique traits. Water is clear and wet, but so are many oils (and muddy water isn't all that clear). Water sustains life, but so do many foods (and water with a few invisible pathogens can kill you). Water freezes when it gets cold, water soaks into wood, water flows downhill, on and on the list grows; but none of these traits, nor any combination of these traits, is both necessary *and* sufficient. No definition devised in the eighteenth century could have captured the true essence of water—the molecule with two hydrogen atoms and one oxygen atom.

By the same token, they argue, scientists in the early twenty-first century are in no position to define life. We have yet to articulate the theoretical underpinnings of biology; we have nothing analogous to the periodic table for living entities. And with only one unambiguous example, cellular life on Earth, we are in no position to lock ourselves into any precise definition. Better, therefore, to keep an open mind and simply describe the characteristics of whatever we find.

I suspect that any universal theory of life will rest, at least in part, on the ideas of emergence. If life arose as a sequence of emergent steps, then each of those steps represents a taxonomically distinct, fundamentally important stage in life's molecular synthesis and organization. Each step deserves its own label.

AN EXPERIMENTAL STRATEGY

Ultimately, the key to defining the progressive stages between nonlife and life lies in experimental studies of relevant chemical systems under plausible geochemical environments. The concept of emergence simplifies this experimental endeavor by reducing an immensely complex historical process to a more comprehensible succession of measurable steps. Each emergent step provides a tempting focus for laboratory experimentation and theoretical modeling.

This nontraditional view of life's definition as a stepwise transition from chemistry to biology is of special relevance to the search for life elsewhere in the universe. It's plausible, for example, that Mars, Europa, and other bodies in our solar system progressed only part way along the path to cellular life. If so, that's crucial to know, at least from NASA's point of view. If each step in life's origin produced distinctive and measurable isotopic, molecular, and structural signatures in its environment, and if such markers can be identified, then these chemical features become observational targets for planned space missions. It's possible, for example, that primitive prebiotic isotopic, molecular, and structural forms are inevitably eaten by more advanced cells and survive as "fossils" only if cellular life never developed in their environs. Thus prebiotic features may serve as extraterrestrial "abiomarkers"—clear evidence that molecular organization and evolution never progressed beyond a certain precellular stage. As scientists search for life elsewhere in the universe, they may be able to characterize extraterrestrial environments according to their degree of emergence along this multistep path.

Consider Saturn's recently visited moon Titan as a choice example. Cloud-enshrouded Titan possesses an atmosphere one-and-a-half times thicker than Earth's and is rich in methane and ammonia. Organic molecules, which color the atmosphere a hazy orange, rain onto the surface to form thick accumulations of organic gunk. Lakes of methane and ethane occur side-by-side with frozen expanses of rock-hard water ice, though conditions are generally much too cold for liquid water or significant chemical progress toward life.

From time to time, however, the impact of a large comet or asteroid may have melted regions of ice on Titan. For periods of hundreds or even thousands of years, gradually cooling ice-covered lakes might have supported the first chemical steps in the path toward life, only to become frozen again. Such primitive biochemistry, though lost forever on Earth's scavenged surface, might conceivably survive in the deep-freeze of Titan.

But so much for speculation and conjecture. Observations of the living world, coupled with relevant experiments, will illuminate the emergence of life both here on Earth and even elsewhere in our solar system.

3

Looking for Life

Scientists turn reckless and mutter like gamblers who cannot stop betting.

Alan Lightman, *Einstein's Dreams*, 1993

The profound difficulty in crafting an unambiguous definition of what is (or was) alive came into dramatic focus in 1996 with the discovery of supposed cellular fossils in a meteorite from Mars. Of the countless thousands of meteorites that have been collected on Earth's surface, only a precious two dozen or so came to us from Mars. In the 1980s, chemists deduced the distant origins of these rocks from the diagnostic composition of gas trapped inside them—gas that matches perfectly the known idiosyncrasies of the Martian atmosphere. Theorists maintained that giant asteroid impacts on Mars could easily have hurled rocky debris into orbit around the Sun. And while the Sun and Jupiter, the two most massive objects in our solar system, eventually (often after millions of years) sweep up most of that Martian detritus, a tiny fraction of the rubble inevitably finds its way to Earth. With the discovery of Martian meteorites, scientists could, for the first time, investigate actual pieces of another planet.

Naturally, these nondescript chunks of dark-colored rock are highly prized and receive the closest examination by earthbound scientists. Most of them are hunks of ancient igneous formations—material formed from once-molten rock near the Martian surface. We expect such meteorites to be devoid of life. But one Mars meteorite proved strikingly different from the others, and it naturally attracted extra close scrutiny. Collected in 1984 from the Allan Hills region of Antarctica (hence its now famous designation, ALH84001), this meteorite held a suite of minerals that suggested to some scientists the possibility of ancient interactions with liquid water.

A team of biologists, planetary scientists, and meteorite experts led by NASA's David McKay subjected pieces of the two-pound rock to a battery of analytical tests. They probed the meteorite with X-rays, lasers, gamma rays, and beams of electrons, recording characteristics as small as a billionth of an inch across. No one had ever expected to find hard evidence for Martian life, but even a hint of freely flowing water on Mars would constitute a major discovery. Yet gradually, as the data piled up, McKay and his colleagues began to believe that they had found the smoking gun for Martian life.

LIFE ON MARS: THE ALLAN HILLS STORY

On August 7, 1996, the Allan Hills team publicly claimed the discovery of tiny elongated objects that were once alive. "LIFE ON MARS!" screamed the headlines, while the prestigious periodical *Science* published an article with the equally giddy title (at least for a scientific journal), "Search for Past Life on Mars: Possible relic biogenic activity in Martian meteorite ALH84001." President Clinton got into the act by holding a national press conference, during which he basked in the reflected glory of NASA's triumph.

McKay and his eight co-workers pointed to five separate types of data, which they presented point by point like a zealous prosecutor at a jury trial. Point number one: The meteorite was found to contain a suite of organic molecules, including carbon-based compounds called PAHs (polycyclic aromatic hydrocarbons). These sturdy, long-lasting molecules, which feature interlocking rings of six carbon atoms, often arise when once-living cells are subjected to high temperature. Since carbon is the key element of life as we know it, its presence in ALH84001, which distinguished that specimen from the other Martian meteorites, was of extraordinary significance.

Point two: The meteorite held microscopic globules of carbonate minerals, similar to those that make the graceful formations on the walls of caves on Earth. Such carbonates are often deposited through the action of liquid water passing through a system of cracks and fissures. Liquid water is the presumptive medium of all cells and thus a necessary condition for life. What's more, their tiny structures, about a ten-thousandth of an inch in diameter, reminded some observers of minerals precipitated by microbes on Earth.

The third and fourth points relied on sophisticated analytical tools. The NASA team used an electron microscope to discover and characterize two iron-bearing minerals, an iron sulfide called pyrrhotite and an iron oxide called magnetite. Of particular interest were the curious chainlike arrays of minuscule magnetite crystals. Magnetite is a magnetic mineral found in abundance in rocks of all types, but the perfect shape of these alien crystals and their unusual chemical purity, coupled with their distinctive linear arrangements, seemed unlike anything ever seen except in a few remarkable types of bacteria. These "magnetotactic" microbes tend to live in thin layers of sediment where chemical conditions change rapidly with depth, and they use their internal magnets to distinguish "up" from "down," by sensing the inclination of Earth's magnetic field. So sensitive are these organisms to their vertical position that magnetotactic bacteria from the Northern Hemisphere move in the wrong direction and die when placed in Southern Hemisphere soils, where magnetic "up" and "down" are reversed. The NASA scientists claimed that no known inorganic process could have produced such an ordered crystalline array.

Finally, the fifth point: ALH84001 holds myriad tiny sausage-shaped objects reminiscent of some species of terrestrial bacteria. Though much smaller than any known Earthly microbes, these suggestive forms provided the public with its most convincing evidence for Mars life. Hundreds of newspapers and magazines reproduced the NASA electron microscope images with captions identifying them as "Martian microbes."

The main text of the McKay et al. six-page article in *Science* conveyed a sober and reasoned discussion of their findings, and they acknowledged that no single line of evidence was enough to trumpet the discovery of alien life. But the concluding sentence shifted tone and pushed the limits of most readers' credibility: "Although there are alternative explanations for each of these phenomena taken individually, when they are considered collectively, particularly in view of their spatial association, we conclude that they are evidence for primitive life on early Mars."

To paraphrase the late Carl Sagan, extraordinary claims require extraordinary proof. Predictably, controversy exploded around the

NASA scientists' bold claim. Experts pored over the paper, which was aggressively challenged on every point.

Point number one: PAHs and other carbon molecules litter the cosmos, notably in the interstellar dust that forms comets and asteroids—the raw materials that formed Mars. What's more, such molecules would have formed in abundance by natural chemical processes at or near the primitive surface of Mars. And PAHs are among the most common constituents of pollution on Earth; the meteorite could have become contaminated while sitting on the ice. There's no reason to conclude that these PAHs represent the remains of living cells.

Point two: The carbonate minerals could have formed in many ways other than by circulating water. Carbonates can occur in reactions of rock with carbon dioxide, the most common Martian atmospheric gas. Carbonates commonly grow as alteration products, long after the host rock forms, or directly from melts by igneous processes. Indeed, a number of researchers reanalyzed the minerals and found evidence that they had formed at temperatures well above the boiling point of water.

Skeptical experts also argued that the minute magnetite crystals prove nothing, since they are common constituents of meteorites that bear no possible signs of life. The chainlike arrays of exceptionally pure magnetite crystals are unusual, to be sure, but most observers feel that magnetite grains are insufficient by themselves to prove the existence of Martian life. Magnetotactic bacteria, furthermore, would have required a moderately strong Martian magnetic field—perhaps stronger than geophysical evidence suggests.

Finally, the purported fossil microbes are too small—an order of magnitude smaller than any known Earthly bacteria. In fact, they are so small that they could contain no more than a few hundred biomolecules—not nearly enough for a living cell. And there's no reason to characterize them as fossils, since inorganic processes (including sample processing in the lab) are known to produce similar elongated shapes.

The story became even more confused when scientists began examining other meteorites, Martian and otherwise, in the same meticulous detail afforded the Allan Hills specimen. Surprisingly, all meteorites reveal signs of life—Earth life. Meteorites smash into Earth, where our planet's ubiquitous microbes inevitably contaminate them. Almost every meteorite ever found has lain on the ground for periods ranging from several days to many thousands of years. Once found, they are usually handled, breathed on, and otherwise exposed to more contamination. Unless hermetically sealed almost immediately, any meteorite will be compromised. In a matter of months, microbes migrate deep into a meteorite's interior, exploiting every crack and crevice in a search for the chemical potential energy that is stored in the meteorite's minerals. Given such a messy environment, how could anyone ever be sure about ALH84001?

One of the most vocal critics of the Martian claim was UCLA paleontologist J. William Schopf. A leading expert on microfossils and an authority on Earth's most ancient life, Schopf was outraged at what he regarded as the NASA team's shoddy analysis and unwarranted conclusions. At the well-publicized August 1996 NASA press conference to discuss the discovery, Schopf was invited to participate as an objective, dissenting voice. "I was like Daniel in the lion's den," he recalls. Not wanting to publicly denigrate the NASA crowd, he may have pulled his punches in that public forum ("I had tried to be reasonable, even gentle"), but he underscored his criticisms of the NASA work in a scathing addendum to his popular book, *Cradle of Life* (1999). There he attacked the NASA team with a withering analysis, which he intensified by juxtaposing his critique of ALH84001 with stories of the most egregious paleontological blunders of all time. Of the late famed meteorite, he wrote: "The minerals can't prove it. The PAHs can't either. The 'fossils' could—but they don't, and there are good reasons to question whether they are in any way related to life."

Schopf concluded on a more philosophical note: "There are fine lines between what is known, guessed, and hoped for, and because science is done by real people these lines are sometimes crossed. But science is not guessing." Little did he suspect that within a few years those righteous proclamations would come back to haunt him.

EARTH'S OLDEST FOSSILS—
THE SCHOPF–BRASIER CONTROVERSY

The top-down approach to life's origins requires that we ferret out and characterize Earth's most ancient fossil life. Those fragile, fragmentary clues may help us bridge the gulf between geochemistry and biochemistry, and thus deduce key steps in life's emergence.

Fossil microbial life should be vastly easier to detect in Earth's ancient rocks than in the handful of meteoritic fragments from Mars. After all, we can collect tons of specimens, scrutinize their geological setting, and check any critical measurements in many different laboratories. No matter how remote the rocks or treacherous the journey, it's well worth the effort, for Earth's earliest fossils not only provide a glimpse of the size and shape of ancient life but also reveal the timing of life's opening act.

Planet Earth formed about 4.5 billion years ago as a giant, molten, red-hot glowing sphere—the result of the accumulation of countless comets, asteroids, and other cosmic debris. For another few hundreds of millions of years, an incessant meteoritic bombardment pulverized every square inch of Earth's surface. What's more, every few million years an epic impact of an object a hundred kilometers or more across punctuated the steady rain of smaller boulders. Such catastrophic events would have repeatedly vaporized any nascent oceans and blasted much of the primitive atmosphere into space. No imaginable life-form could have survived the hellish onslaught of that so-called Hadean eon.

We don't know exactly when cellular life arose, but the window of opportunity appears to have been surprisingly short. It's almost certain that life could not have persisted before about 4 billion years ago, when the last of the great globe-sterilizing events is estimated to have occurred. It's always possible that life began several times before that, only to be snuffed out by the periodic impact of devastating asteroids. In any case, chemical evidence for life in Earth's oldest known rocks— formations 3.5 to 3.8 billion years old from Greenland, South Africa, and Australia—seem to establish a remarkably ancient lower age limit for life. Such a narrow time window suggests that life's emergence was rapid, at least on a geological timescale.

Paleontologists devote their lives to scrutinizing fragmentary signs of life in rocks. It's not always a glamorous business, mucking about in inhospitable, remote landscapes, but there's always the possibility for making a big splash. Paleontologists, perhaps more than scientists in any other discipline, can generate gripping headlines. Discoveries of history's biggest shark, most massive dinosaur, or oldest human inspire the public imagination. We live in an age of Guinness-style records; we are obsessed with superlatives. One recent report in *USA Today* even trumpeted the discovery of the oldest known fossilized penis in a 400-million-year-old crustacean!

With such a fossil-obsessed press corps, it's little wonder that paleontologist Schopf made the evening news (and *Guinness World Records*) in April 1993 with his announcement in *Science* of the discovery of Earth's oldest fossils ("Microfossils of the Early Archean Apex Chert: New Evidence of the Antiquity of Life"). Schopf claimed to have identified actual single cells, preserved in the 3.465-billion-year-old Apex Chert from the sun-baked northwestern corner of Western Australia. Even more surprising, these cells occurred in filament-like chains strongly reminiscent of those formed by modern photosynthesizing microbes—cells with the relatively advanced chemical capability to harvest sunlight.

As in the subsequent ALH84001 incident, the claims were extraordinary and consequently demanded extraordinary proof. In this case, however, the geological community was generally quick to accept Schopf's assertions, because he had established a reputation as one of the world's leading experts in finding and describing ancient single-celled microbes. Schopf and his students had already catalogued dozens of new microbial species from 2-billion-year-old rocks around the world, while establishing rigorous standards for the cautious identification and conservative reporting of new finds. The latest fossils merely pushed back the record for the world's oldest life a few hundred million years.

A straightforward UCLA protocol had become standard for the maturing field of micropaleontology. Visit Earth's geological formations of the Archean eon (4 billion to 2.5 billion years ago), identify layers of sediment that were deposited in ocean environments, and scour the region for outcrops of distinctive carbon-rich rocks called black chert. Field-workers collect hundreds of pounds of Archean rocks, break off hunks of the most promising specimens, and ship them back to California, where they are sliced into 2 × 3-inch transparent thin sections, a few hundredths of an inch thick.

The research protocol for finding ancient microbes can be exceptionally tedious. Graduate students are coaxed and coerced into spending thousands of hours examining every part of every slide, micron by eye-straining micron. It turns out that black chert isn't really black at all. Illuminated from beneath and viewed in a powerful microscope, thin sections provide a window on the ancient world. The typical cherty matrix is chockablock full of little black blobs and smudges. Most black chert is seemingly barren of life, but once in a while a thin section reveals a host of tiny spheres, disks, rods, and chains—dead ringers for modern bacteria. Schopf was fortunate that in 1986 one especially sharp-eyed and conscientious student, Bonnie Packer, scrutinized the most promising Australian specimens. Most thin sections yielded nothing of interest, but her discovery of unambiguous microfossils in several ancient units led to a prominent publication and set the stage for the Apex controversy.

Appearances can be deceiving. Lots of inorganic processes produce round specks and enigmatic squiggles. It's all too tempting to see what you want to see in an ancient rock. That's why Schopf and his colleagues had developed an arsenal of confirmatory tests. For one thing, size matters. Single-celled organisms can't be too small or too big (though some remarkable ancient single-celled organisms are monsters by modern standards). Even more critical, microbial populations tend to cluster tightly around one preferred size, in contrast to the more random sizes of structures produced by nonbiological processes. Consequently, a statistical analysis of size distributions often accompanied Schopf's papers. Uniformity of shape is another key; no fair photographing one or two suggestively contoured black bits while ignoring a multitude of shapeless blobs. Schopf also demanded rigor in the description of local geologic setting and in the proper dating of his samples. As a result, his work on the Apex Chert was initially accepted; he had established a solid reputation for cautious, conservative science.

But one aspect of Schopf's 1993 study—the claim that some of the microbes were photosynthetic and hence oxygen-producing—remained puzzling. Geochemical evidence from Earth's oldest rocks points to an oxygen-poor atmosphere prior to about 2.2 billion years ago, a time that most researchers identify with the rise of photosynthesis. How could oxygen-producing microbes be present more than a billion years earlier? Nevertheless, within a few years Schopf's claims for the earliest fossils were standard textbook fare; his pictures of Apex fossils had become among the most frequently reproduced of all paleontological images. Schopf himself highlighted the historic findings in *Cradle of Life*. [Plate 2]

Controversy erupted in March 2002, after Oxford paleontologist Martin Brasier and a team of seven British and Australian colleagues conducted a careful reexamination of the original type specimens of the Apex Chert fossils, which had been deposited at the Natural History Museum in London. Brasier employed a microscopic technique called image montage, which allowed him to use sharp images of the original thin sections at many different levels within the rock slice to reveal three-dimensional details that were not previously obvious.

Brasier's microscopic investigation cast the Apex fossils in a new light. Their 3-D structures seemed to differ sharply from those of any known cellular assemblages. In some cases the "filaments" appeared to be more like irregular planes or sheets. In others they branched, a feature never observed with cells. Brasier gave some of the more curious shapes nicknames like "wrong trousers" and "Loch Ness monster." What's more, the thin sections with the most convincing cell-like objects contained numerous additional black shapes that bore no resemblance at all to cells—forms that Schopf must have seen but failed to detail in his *Science* paper.

Further study by Brasier's geological colleagues in Australia pointed to other discrepancies. Schopf had visited the site only briefly and, based on the linear character of the outcrop, reported a classic layered sedimentary sequence with the black chert lying between other layers—a typical ocean-floor scenario. But after detailed field mapping of the site, Australian geologists Martin van Kranendonk and John Lindsay realized that the geological setting of the Apex Chert was much more complex than the simple layered formation Schopf had described. Indeed, the Apex Chert formed at the site of significant hydrothermal activity, where hot volcanic fluids circulated through cracks and fissures. According to their reinterpretation, the black chert formed as a consequence of fluids circulating through this dynamic system as part of a cross-cutting vein. Given this relationship, with the vein of chert cutting across older rocks, the exact age of the Apex Chert was called into question. More damning still, the hydrothermal setting suggested that the chert formed at temperatures far above the permissible limits for life.

Brasier et al. challenged Schopf's claims in an article titled "Questioning the evidence for Earth's oldest fossils," published in 2002 in the widely read journal *Nature*. Their bold conclusion: "We reinterpret the purported microfossil-like structure as secondary artifacts." The article was a very public attack on Schopf's credibility.

In an unusual move, the editors of *Nature* had delayed the Brasier et al. article for more than a year, to allow Schopf time to prepare a rebuttal, "Laser-Raman imagery of Earth's earliest fossils." The two conflicting articles appeared back-to-back in the March 7, 2002, issue. An accompanying "News and Views" analysis by *Nature* staffer Henry Gee emphasized the irony of Schopf's predicament.

Seldom has a scientific debate held such high drama. Schopf had made his reputation in part by staking claim to Earth's oldest life, while cutting no slack for the questionable claims of others. More than any other scientist, he had thrown cold water on the NASA pronouncement of life on Mars. He reveled in reminding the public of past paleontological follies. No wonder then that science journalists were quick to highlight the controversy: "CRADLE OF LIFE OR CAULDRON OF CRUD?" one news headline asked.

This debate came to a head on April 9, 2002, at the second biennial NASA Astrobiology Science Conference, with Schopf and Brasier squaring off like graying, bespectacled wrestlers. The entertaining spectacle took place deep inside the gargantuan antique dirigible hanger of Moffett Field, 30 miles south of San Francisco, which is home to the NASA Ames Research Center. A sturdy lectern embossed with the NASA logo stood on the stage, to the left of a large projection screen about 12-feet square. Both speakers were seated on the stage, before a rapt audience of several hundred scientists.

Schopf spoke first. A flamboyant presenter even under the calmest of circumstances, Bill Schopf was fighting to preserve his scientific reputation. Barely controlling his anger, his voice booming, he lectured Brasier as if the Englishman were a recalcitrant schoolchild. Step by step, in a talk rich in withering rhetorical questions and exaggerated dramatic pauses, he reviewed the dozen or so necessary and sufficient criteria to establish the authenticity of ancient fossil cells. Step by step, he provided the data to back up his Apex claim, though he did soften his assertion that the microbes were oxygen-producing cyanobacteria.

After 15 minutes or so, the moderator gestured that Schopf's allotted time was almost up. Like a magician pulling a rabbit out of a hat, Schopf concluded by displaying new analytical data that he claimed would prove his case once and for all. The smudgey black Apex Chert "fossils" are composed principally of carbon, the essential element of life. Carbon concentrations may arise by both biological and nonbiological processes, so carbon in and of itself is not diagnostic of life. However, Schopf claimed, there is a difference: The carbon remains of fossil cells are less perfectly ordered than crystalline carbon deposited as a lifeless mineral. The degree of crystallinity, furthermore, can be revealed by the established technique of Raman spectroscopy. Schopf grandly presented a suite of Raman spectra: Indeed, sharp spiky peaks characteristic of inorganic carbon stood in sharp contrast to the "obviously biological" broad humps in the Raman spectra from the Apex Chert. Schopf concluded by summing up all the evidence he had mustered: "If it fits with all other evidence of life, well follks, most likely it's life." [Plate 3]

Brasier gently ascended the stage and began his rebuttal with a dismissive putdown of his rival's presentation: "Well, thank you, Bill, for a truly hydrothermal performance. More heat than light, perhaps." In soft-spoken Oxford English, the tone in sharp contrast to what had come before, he began to cast doubt on Schopf's case. The most damning evidence were the fossils themselves. With the right lighting, field of view, and level of focus, the Apex features do look like strings of cells. The size is right, the shape more than a little convincing, and there are even regularly spaced dark divisions that look like cell walls. But raise or lower the focus slightly, or shift to another field of view, and doubts arise. What are all those shapeless black blobs next to the "fossil?" How can that supposed straight chain of cells suddenly branch like a "Y"?

As Brasier warmed to his task, an agitated Schopf stood up and began to pace distractingly a dozen feet behind the podium. Back and forth he walked, hunched over, hands clasped firmly behind his back—a tense backdrop to Brasier's staid delivery.

Ignoring these diversionary tactics, Brasier fired salvo after salvo. Schopf had the geology all wrong, he claimed. A new detailed geological map of the Apex area suggested that the black chert filled a cross-cutting vein—evidence that the chert had formed much later than the surrounding rocks, through the agency of hot circulating water. He outlined chemical experiments that produced cell-like chains of precipitates in a purely inorganic setting—nonliving structures similar to the supposed Apex fossils form with ease under the right chemical circumstances. He demonstrated how carbon-rich deposits might have formed nonbiologically through a familiar industrial process called the Fischer–Tropsch synthesis. He even showed his own Raman spectroscopic data of inorganic carbon that had the same broad features as the purported biological carbon of Schopf's fossils.

As Brasier calmly outlined his arguments, the scene on stage shifted from awkwardly tense to utterly bizarre. We watched amazed as Schopf paced forward to a position just a few feet to the right of the speaker's podium. He leaned sharply toward Brasier and seemed to glare, his eyes boring holes in the unperturbed speaker. After a few seconds, Schopf retreated to the back of the stage, only to return and stare again. Perhaps Schopf was just trying to hear the soft-spoken Brasier in the echoing hall, but the audience was transfixed by the scene.

The two presentations ended in due course and, after an extended period for audience questions and comments, the session concluded. Many of us breathed a sigh of relief that no blows had been exchanged, and then we tried to figure out who won. We all knew, of course, that science isn't about winning. The black smudges in the Apex Chert were either the remains of ancient microbes or they weren't. Eventually, we all assumed, the truth would be found out. A debate like the Schopf–Brasier bout did little but outline the problem and establish our collective state of ignorance. Still, we wondered: Who won?

To be sure, Schopf's intense delivery and unconventional antics hadn't won him any points among my acquaintances. Many scientists were also struck by the sudden softening of his previous claims that his fossils were cyanobacteria. Such waffling undermined a decade of confident, highly public interpretations. But Schopf is also a fine scientist with a long track record; and his systematic point-by-point analysis of the fossils, however quirky in its delivery, appeared both logical and persuasive.

Brasier's cool detachment, by contrast, seemed calculated to provide a veneer of objectivity, yet that very lack of passion and intensity may have cost him some points. So much of the Apex story relied on interpretation of fuzzy objects in a fuzzier context. As doubtful as Schopf's claims might be, it was equally difficult to *disprove* any biological activity by pointing to irregular black shapes. We have no way of knowing what 3.5 billion years of decay might have done to ancient microbes, and in many ways Brasier's arguments were just as subjective as Schopf's. Rather than providing the audience with the smoking gun that would thoroughly discredit Schopf, Brasier seemed merely to have raised a number of serious doubts—knotty technical issues that deserved further study.

Meanwhile, paleontologists around the world, Schopf and Brasier included, keep searching thin sections of ancient rocks in hopes of finding Earth's earliest fossils.

If there is a moral to the Allan Hills meteorite and Apex Chert controversies, it is that unambiguous identification of ancient life from microscopic structures is fraught with difficulty. Tiny rods and spheres are not always useful indicators of biology. The older the rock, the more difficult the interpretation of such vague features becomes. If fossils are to provide any clues about life's ancient emergence, then we have to look beyond microscopic structures to the tiniest fossils of all.

4

Earth's Smallest Fossils

Millions of brutal years of burial and resurfacing, akin to
repeated pressure cooking, permitted very few fossilized cells to
survive. . . . Often geologists must instead rely on other signs of
life, or biosignatures—including rather subtle ones, such as
smudges of carbon with skewed chemical compositions unique
to biology.

Sarah Simpson, 2004

Even as the Schopf–Brasier battle raged, a small cadre of less publicized researchers labored to craft a convincing case for fossils even
more ancient than Apex. This new breed of paleontologist doesn't depend on questionable black blobs. They probe rocks for fossils far
smaller than microscopic cell-like spheres or segmented filaments. Remarkably, the fossils they seek consist of the very atoms and molecules
of once-living organisms.

When a cell dies, its vital chemical structures quickly fragment and
decay. Almost always the essential atoms of biochemistry—carbon,
hydrogen, oxygen, nitrogen, and more—disperse and return to the environment. Earth's vast but nevertheless finite reservoirs of life-sustaining atoms play their parts over and over and over again. Most of
the atoms in your body were once part of mastodons, dinosaurs, trilo-
bites, even the earliest living cells. Take a moment to look at the palm
of your hand and imagine the fantastic yet unknowable histories of its
countless trillions of atoms. Earth's biosphere is the ultimate recycling
machine.

Atoms almost always recycle, but once in a great while, under an
unusual concatenation of geological circumstances, a dying organism
will find itself encased in an impermeable rock tomb. If a worm is

swept away and buried in a sudden mudslide, if a colony of deep-sea microbes solidifies in chert, if a winged insect dies ensnared in sticky tree sap, then it's just possible that some of the organism's original atoms and molecules will become trapped as well. Such a trapped fossil animal or microbe may persist through eons in its original form, or it may decay to a shapeless dark splotch. Nevertheless, its hermetically sealed atoms and molecules are the remains of past life, so they qualify as fossils just as legitimate as the most elegant coiled ammonite or massive dinosaur.

FOSSIL ATOMS

It's an amazing feeling to hold a 3-billion-year-old rock that once teemed with living organisms—a sample that contains the very atoms and molecules of cells from the dawn of life. Such rare and precious samples demand a new approach to the study of fossils; traditional descriptive paleontology must morph into analytical chemistry.

A casual conversation during the summer of 1997 with longtime friend Andrew Knoll, professor of paleontology at Harvard University, led me into this fascinating field. Andy and I were attending a Gordon Research Conference on the origin of life, held at New England College in Henniker, New Hampshire. He's an engaging, articulate, and friendly speaker, and the author of richly illustrated articles and lectures on the diversity of microbial fossils in Earth's oldest rocks—presentations that opened a new world to me.

When most people hear the word "fossil," they think of the bones of a dagger-toothed *Tyrannosaurus rex* or the spiny shell of a trilobite—hard parts that survive the rigors of decay and burial. By contrast, the soft cellular tissues of animals, plants, and microbes almost always rot away without a trace. Only occasionally will an organism die and be buried in rock fast enough to preserve cellular detail. For a micropaleontologist like Andy Knoll, whose specialty is ancient microbes, those rare cellular fossils provide the raw material for a career in science.

The very earliest fossil cells are nondescript objects and difficult to identify, but geologists have documented dozens of localities with numerous clearly identifiable microfossils dating from about 2.9 billion years on. Distinctive bumpy rods and symmetrically spiky spheres, chainlike filaments of repeated rectangles, and curious corkscrew spirals form a panorama of primitive life. Andy's work surveys the saga of life's evolution, culminating in the first enigmatic multicellular organisms about a billion years ago.

Throughout our conversations at the Gordon conference, I was struck by the fact that many ancient microscopic fossil forms are preserved in black chert or shale—impermeable rocks that have the potential to preserve chemical traces of the original bacteria. Over a beer, I asked Andy if paleontologists ever analyzed their microfossils with the kind of machines that we mineralogists routinely employed to characterize the atoms and isotopes of our samples. He shook his head and admitted that, while there had been a few pioneering studies, most paleontologists worried almost exclusively about the sizes and shapes, not the chemistry, of their bugs. Then came his deceptively innocent question: "Do you want to collaborate? I've got a couple of students with really interesting samples. . . ."

My own research on life's emergence had to that point focused on bottom-up chemical experiments, trying to synthesize life's molecular building blocks, but the top-down approach also has great appeal. I've always loved fossils and was more than happy to associate myself with a real paleontological pro, albeit in a modest support capacity. Agreeing to the offer, I immediately envisioned an arsenal of microanalytical tools that might be brought to bear on the problem. Our conversation soon turned to technical details: the number of samples, their size, the degree of chemical alteration, and more.

The first samples arrived at Carnegie's Geophysical Laboratory from Harvard within a few months, and many more followed. A suite of 400-million-year-old plants from Canada, slices of ancient black soils from Australia, 3-billion-year-old microbial mats from South Africa, bizarre spiky spores from a billion-year-old Chinese formation—wonderful fossils holding some of the secrets of life's past. Our Carnegie team quickly confirmed that ancient fossils have the potential to provide three important types of microanalytical data: chemical elements, isotopes, and molecules. Of the three, the composition of chemical elements is arguably the easiest to measure.

The mineralogist's tool of choice for analyzing chemical elements is the electron microprobe, a costly but indispensable piece of hardware in geology departments around the world. The machine works by firing a narrowly focused beam of electrons at a highly polished piece of rock, typically an inch or so across. The energetic electron beam excites the rock's atoms, which in turn emit a spray of X-rays. It turns out that every element of the periodic table produces its own slightly different suite of X-rays of different wavelengths. The task, then, is to capture these X-rays and measure their diagnostic wavelengths.

Microprobe analysis of most elements has become routine and automated. The machine is a workhorse, operating 24 hours a day for analyses of silicon, magnesium, iron, and other rock-forming elements. But the lightest elements, including the one of greatest interest to us— the key biochemical element carbon—pose a severe analytical challenge. Lighter elements tend to be rather inefficient at producing X-rays, while the relatively few X-rays that are produced have rather low energies. Both of these factors complicate carbon analysis. What's more, we routinely use carbon to coat rock samples prior to probing, in order to make them electrically conductive. The coat is essential to prevent the sample from building up an electric charge while being bombarded by electrons, but it can mask any carbon in our samples. In short, the Lab's usual microprobe procedures wouldn't work. We'd have to devise new protocols.

The Geophysical Lab's deviser of protocols is Christos Hadidiacos, microprobe jockey extraordinaire. For more than 30 years, Chris has maintained and upgraded the Lab's electron microprobe. He knows every trick in the book and constantly invents new ones to push the limits of analysis. Complex circuit diagrams, cryptic numerical tables, and other papers decorate his office, while technical manuals and electronics catalogs fill his bookcases and rows of volleyball trophies line the shelves above.

I explained the carbon analytical problem to him—one he had never faced before. The Geophysical Laboratory probe had been used almost constantly for decades but almost always to study rocks and experimental run products—samples with mostly heavier elements. But it took Chris less than a minute to figure out a possible fix. "We can try raising the current," he said. "That might work." He began asking detailed questions about the expected amount of carbon and what other elements might be present. Then a frown. "We're not going to be able to use a carbon coat, are we?"

I shook my head, knowing that this was a potential deal killer. We needed an electrically conductive coating, and that meant a metal. But a carbon coat would mask our fossils, and most metals absorb X-rays so efficiently that we'd never see the low-energy carbon X-ray signal. "Any ideas?" I asked.

Again, it took him less than a minute. "Maybe we could use aluminum. Just vaporize a bit of aluminum foil." He was smiling again, pleased at the simplicity of his solution. Aluminum, element 15, should be light enough itself to allow most of the carbon signal through. It was definitely worth a try.

I handed over the first of my fossil specimens, a pair of 2×3-inch rectangular thin sections of 400-million-year-old plant fossils from Rhynie, Scotland, a classic chert locality. The samples had been collected many years ago as isolated flinty boulders in old stone walls; no rock outcrop has ever been found. The precious thin sections arrived courtesy of Kevin Boyce, a bright-eyed, soft-spoken grad student in Andy Knoll's group. Kevin had trolled through Harvard's somewhat neglected paleobotany collection as part of his thesis work on the evolution of leaves. He hoped that the Rhynie samples, which preserve cellular structures of some of the oldest known land plants, might reveal clues about the chemical evolution of plants.

The first step was to apply the thin aluminum coating. Our antiquated but serviceable vacuum coating system consists of a well-worn metal housing about the size of a washing machine with vacuum pumps and hoses arranged inside. A 4-inch-square platform with wire electrodes sits on top, while vacuum gauges and control valves project from the side.

Chris snipped a 1-centimeter-square piece of aluminum foil, crumpled it up and placed it in a small wire basket attached to the electrodes. He arranged the two Rhynie chert thin sections on the metal platform, and then lowered a 2-foot-tall dome-topped bell jar so that its rubber-lined base made an airtight seal. The old pumping system labored for a quarter of an hour to achieve the desired vacuum, but then it took only a fraction of a second to apply an electric current and vaporize the aluminum foil. Aluminum atoms flew off in all directions, coating everything inside the bell jar, including our samples. Chris released the vacuum, raised the bell jar, and the fossils were ready for the probe.

The Geophysical Lab electron microprobe is an awkward-looking tabletop machine that sits in its own small room. [Plate 3] Two chairs flank the workbench, which is dominated by a 4-foot-high cylindrical tower that looks like a model of some futuristic fortification. The tower houses the electron gun, the heart of the probe. At the tower's top, a coiled tungsten filament generates electrons; a series of ring-shaped electromagnets focus the electrons into a narrow beam as they accelerate downward onto the sample. The base of the tower is cluttered with five boxlike attachments, called spectrometers, that measure X-rays, plus various vacuum lines, power cables, and viewing ports.

A curious combination of instrument panels controls the electron gun hardware and X-ray detectors. To the right, a computer monitor displays all the machine's vital statistics—beam current, spectrometer settings, sample position, and more. In sharp contrast, a 1980s-vintage slant-front console, sporting two antiquated 6-inch black-and-white video screens and more than a dozen plastic knobs reminiscent of a classic *Star Trek* set, dominates the central table. Like an old house that has undergone decades of renovation, the Geophysical Lab probe has been through a lot of upgrades.

I carefully secured one of the thin fossil sections into a shiny metal sample holder, closed the sample port, and waited a couple of minutes for the machine to achieve the high vacuum necessary to stabilize the electron beam. Meanwhile, Chris fiddled with the computer controls, raising the electron current to about ten times its normal settings. It took him a few moments to center and focus the intense beam onto a carbon-rich portion of our sample. We were about to discover whether or not we could detect fossil carbon atoms.

It worked! A carbon signal of 600 X-ray counts per second stood out sharply from the 30-count-per-second background. We were in business. It was a simple matter to select half a dozen areas, each about a fiftieth of an inch square, to map. Slowly but surely, the microprobe beam scanned across the sample, measuring the carbon concentration point by point. It took about 3 hours to produce one map. We put the probe on automatic, happy with our rapid progress, and headed outside to the Lab's sand court for our afternoon game of volleyball.

The analytical procedures took a bit of tweaking. The initial aluminum coatings were too thick, the beam settings not quite optimal. But within a few weeks, we were producing a steady stream of colorful maps; regions rich in carbon atoms from ancient life-forms stood out boldly against the carbon-poor fossil matrix. Cellular features less than a ten-thousandth of an inch across were clearly visible. Armed with these maps, Kevin Boyce, with his sophisticated botanical eye, was able to describe and interpret cellular detail never previously seen. [Plate 3]

Making these carbon maps, watching the fine details emerge, is great fun. Each map is formed from a two-dimensional array of point analyses, just like the pixels on your computer screen. We typically employ a quick 400×400-point array for reconnaissance, while slower 500×500-point arrays yield beautifully detailed maps with colors representing the concentration of carbon—red for the highest carbon content, followed by orange, yellow, and the other spectral colors. We play with map colors like a high-tech video game to heighten the contrast and highlight features of special interest.

Maps of the distribution of fossil carbon atoms can be dramatic and surprising as well as beautiful, revealing subtle cellular-scale details not previously recognized. In a sense, though, this analytical effort is little more than an extension of past morphological studies of fossil size and shape—a slightly more elaborate way to image the specimens. These carbon-rich fossils preserve far more information than just the chemical elements that make them up. That's why, for more than three decades, geologists have examined ancient life for fossil isotopes.

FOSSIL ISOTOPES

The fascinating discipline of atomic-scale paleontology has blossomed primarily because all living cells perform a wonderful repertoire of distinctive chemical tricks. Life transforms any collection of its constituent atoms in subtle and surprising ways. Carbon atoms, for example, come in two common varieties—isotopes dubbed carbon-12 and carbon-13. Every carbon atom has exactly six massive positively charged particles called protons in its nucleus; that atomic number, 6, is the chemical definition of "carbon." The distinction between the two common carbon isotopes lies in the number of neutrons, a second kind of massive particle that also resides in the atomic nucleus. Carbon-12 has six neutrons, while carbon-13 has seven.

The number of neutrons has no bearing whatsoever on carbon's chemical behavior. You could live equally well on a pure carbon-12 diet or a pure carbon-13 diet. But there is one important physical difference: Carbon-13, with its extra neutron, is about 8 percent more massive than carbon-12. (An average-sized person whose cells were made entirely with carbon-13 atoms would weigh about 2 pounds more than the same person made entirely with carbon-12 atoms.) As a consequence of this small mass difference, carbon-13 atoms are also a little more sluggish than carbon-12 atoms when taking part in some chemical reactions. So when living cells process carbon-bearing food, they become slightly enriched in the lighter isotope, carbon-12. That characteristic isotopic signature of life can be preserved for billions of years in rock.

Analytical studies of countless carbon-bearing rocks reveal a sharp dichotomy. Most of Earth's carbon is locked into mineral deposits, notably the abundant carbonates that adorn the landscape with bold limestone cliffs and dissolve to open sublime limestone caverns. Worldwide, this mineral-locked carbon has a well-defined uniform ratio of carbon-12 to carbon-13 of about 99:1—the standard reference value, which is designated as 0. By contrast, living cells are invariably isotopically lighter, with a higher proportion of carbon-12 than in limestone. This difference between limestone and life arises from chemical reactions in cells, which more readily incorporate the lighter carbon isotope.

On the geochemist's peculiar scale, a 1 percent deficiency of carbon-13 relative to standard limestone is called "–10 per mil," a 2 percent drop "–20 per mil," and so on. Such "light" carbon in a rock sample thus carries a negative number value and with it a strong presumption of biological activity. Thousands of ancient fossil specimens, from mammoth bones (about –21) to the 3.1-billion-year-old microbial mats in South African sandstones and the fossils in Western Australian chert (between –25 and –27), bear out this simple relationship. So does fossil coal, the transformed remains of 300-million-year-old swamp life, which typically ranges from –24 to –25. Carbon isotope studies of soft-bodied fossils from the Burgess Shale, a 540-million-year-old British Columbian locality, made famous by Stephen Jay Gould's *Wonderful Life*, display a similarly narrow range of values between –25 and –27. The conclusion: If a rock holds an ancient inventory of carbon atoms from once-living cells, then the carbon invariably will be light, even if all morphological signs of life are gone.

The most ancient microbial samples, which often consist of black, carbon-rich splotches in limestone, shale, or other sedimentary rock, have received special attention in recent years. Dozens of studies on billion-year-old rocks from Africa, Australia, Europe, and North

America reveal consistently negative carbon isotope values, but they also point to a significant scatter among the dozens of known microbial fossils more than a billion years old. Photosynthetic microbes, which live on sunlight, tend to lie in the −20 to −30 range. These organisms have dominated the fossil record since about 2 billion years ago, when Earth's atmosphere became oxygen-rich. However, many types of more primitive single-celled organisms that live off the Earth's chemical energy are much more efficient at concentrating the light carbon isotope, carbon-12. Values as low as −50 have been found in 3.8-billion-year-old sediments. While these differences help paleontologists interpret the varied lifestyles of ancient microbes, all unambiguous cellular fossils contain some proportion of light carbon.

It's amazing how nerve-wracking waiting for a machine to produce a single isotope value can be. I was given my most memorable carbon-rich sample in the summer of 2002, during a lecture tour to Australia. A side trip to Sydney's northwestern suburbs brought me to the campus of Macquarie University. There, in beautiful green landscaped grounds, is the home of the Australian Centre for Astrobiology, whose director is paleontologist Malcolm Walter.

Walter's work on ancient Australian microfossils was well known to me, and it was a delight to meet him. He welcomed me with a strong handshake, keen eyes, and ready smile and gave me a quick tour of the facility, a tweed jacket his only protection from the mild Sydney winter. My research was unknown to him, though he was well aware of the well-funded NASA Astrobiology Institute, of which I was part. He listened as I described our various research projects, but his ears really perked up when I recounted my collaborations at Carnegie with Andy Knoll and his students. I outlined our procedures for mapping carbon atoms and described some recent carbon-isotope results.

"You might be interested in this," he said, handing me a small cloth bag containing several black rock fragments. "It's Strelley Pool Chert, a new find from Trendall in Western Australia. It's almost as old as the Apex."

The chance to hold, much less study, one of Earth's oldest rocks is a rare privilege. My response was rather pointed and less than subtle: "We'd be happy to do the carbon work, if you'd like. I could do it next week when I get back." Few laboratories in Australia had the facilities to analyze the ancient rock, while Carnegie was set up and ready to go.

A brief cloud of concern seemed to pass across Walter's face, but he hesitated only a moment before extracting a thumbnail-sized sample, a tiny fraction of the valuable hoard. "Perhaps you could have a look at this," he said, and told me that he was eager to find out as soon as possible whether or not the rock's carbon was light. I slipped it into my pocket, hardly believing my good fortune in having acquired a piece of Earth's earliest history. The sample would be a top priority on my return, I assured him, and any data would be his to announce or publish as he wished. Our conversation shifted to less scientific matters: the constant stress of raising money for the Institute, and the glories of his sheep farm in the countryside, where he spends his weekends.

For the rest of that Australian trip, the tiny, 3-ounce sample weighed heavily on my mind, and it rose right to the top of a long "to do" list on my return. When I need a carbon-isotope analysis at the Carnegie Institution, I turn to Marilyn Fogel, a biologist who has amassed an impressive arsenal of analytical hardware. Marilyn and her group study the cycling of elements through ecosystems, and there's no better way to track an element than with isotopes. Carbon and nitrogen isotopes, both of which get progressively lighter as you move up through the food chain, are her specialty. Fogel's field areas tend to be exotic: the crocodile-infested mangrove swamps of Belize, the parched outback of Australia, the boiling springs of Yellowstone Park. Her growing scientific reputation and thoughtful mentoring style attract a steady stream of postdocs and visitors.

I grabbed the stone and headed down the hallway mid-morning. After the requisite niceties I got directly to the point: "I met Malcolm Walter in Sydney." I reached into my pocket and handed her the piece of black chert. "Take a look at this."

"Apex?" she asked. We all had been following the Schopf controversy, and she knew that Apex Chert samples were pretty hard to come by.

"Nope, this is apparently new. Same area but different. He'd like a carbon-isotope value."

Marilyn doesn't generally betray excitement, but she immediately knew the significance. This black fragment was one of the oldest rocks on Earth. "Wow, that's pretty neat!" She turned the object over in her hands. "Yes, I think the machine is available this afternoon." She paused. "Why don't you come by around two."

I passed the time by breaking off a small piece of the chert, crushing it in a mortar, immersing the powder in oil, and peering at the tiny glassy shards through a powerful microscope. How curious it was—unlike any rock I'd seen before. Myriad tiny black specks, each a few ten-thousandths of an inch across, clouded the otherwise clear, colorless chert matrix. Unlike typical fossil microbes, which tend to occur in clumps and filaments, these dots were uniformly dispersed. They certainly looked like carbon, but were they cells? Would they show a negative isotopic signature?

At two o'clock I showed up at Marilyn Fogel's lab, as arranged. Isotope experts rely on mass spectrometry, the experimental technique of choice for measuring a sample's isotopic ratio. Marilyn's mass spectrometer for carbon-isotope analyses sits in one corner of a 20 × 40-foot room crowded with scientific hardware. Little space is wasted, and you have to exercise care not to bump into sensitive hardware when squeezing between the various experimental stations. These days, mass spectrometers tend to be highly automated and incredibly precise machines, though they still require meticulous maintenance and rigorous standardization procedures to yield reliable results. With a machine like Fogel's, carbon-isotope analyses are relatively straightforward. [Plate 4]

The mass spectrometer is a conceptually elegant analytical tool grounded on two of the great physical laws of nature. Newton's second law of motion, $F = ma$ (force equals mass times acceleration), enables the separation of two atoms of different mass. As noted, carbon-12 and carbon-13 differ by about 8 percent in their mass, so if the two isotopes are subjected to an identical force, then the carbon-12 atom will accelerate about 8 percent faster than the carbon-13 atom. Mass spectrometers accomplish this acceleration by applying a second fundamental law, related to electricity and magnetism: Magnetic fields exert forces on electrically charged particles. The analytical technique is to ionize the carbon atoms: Strip an electron from each—by zapping them with a laser, for example—to yield carbon atoms with a positive electric charge, then subject the ionized atoms to a powerful magnetic field. In some mass spectrometers, a massive horseshoe-shaped magnet bends the stream of carbon atoms (bending is a kind of acceleration, as you discover when you ride a twisting roller coaster). Carbon-12 atoms curve in a tighter arc than carbon-13, so the beam of carbon atoms separates into two. Two detectors placed side-by-side measure the relative amounts of the two isotopes. Alternatively, sensitive electronic detectors measure the time of each ion's flight:

carbon-12 atoms arrive at a target a fraction of a second before carbon-13 atoms.

Fogel's mass spectrometer works best with a small powdered sample, so I had crushed and ground a chip of chert to specifications. Rocks are generally a lot easier to prepare than plant and animal tissues, which must be freeze-dried first. In duplicate, I carefully weighed about a milligram of rock powder and tightly wrapped the powder into a tiny ball of inert tin metal foil. Marilyn inserted a series of well-documented carbon-isotope reference standards along with my samples, each a crumpled metal sphere about the size of a BB, into ports of the mass spectrometer's automated sample holder. Then a computer control system took over and we had to sit back and wait for that one tantalizing number.

It takes only a few minutes per sample, but it seemed longer. The standards always come first, of course; we have to make sure everything is working properly. Finally, the machine spewed out a single printed sheet of white paper, crammed with columns of numbers. One number at the bottom was the key: −25.7 ± 0.5. The carbon was light—just what we've learned to expect from ancient microbes! The analysis also indicated that about a tenth of a percent of the chert's mass was carbon. The duplicate run soon followed: −25.9—satisfyingly consistent results.

But even as we saw these enticing numbers, a nagging doubt remained about the biological origin of the carbon. The oddly uniform distribution of black specks in the chert looked nothing like fossil forms. Indeed, the uniform spacing suggested a more chemical process—a segregation of carbon from chert as oil drops separate from water. Might there be nonbiological pathways to such a light carbon signature? All known life-forms have a negative isotope signature, but does a negative isotope value provide unambiguous proof of life? We were convinced that the Australian chert had a fascinating story to tell.

I e-mailed Malcolm Walters right away with the exciting data and a quick analysis of their possible significance. It would take a bit more work, I thought, but these were certainly publishable results. His reply came slowly, and with a disappointing surprise. Please stop working on the samples, he asked. Evidently, Walter's Australian colleague, paleontologist Roger Summons, had been promised the chance to analyze the new find. Summons, a pioneer at extracting biomolecules from old rocks, had recently accepted a professorship at MIT and had already lined up a graduate student to do the work. It wouldn't do for our efforts to undermine that thesis project.

A deal's a deal. I put aside the chert and went back to the seemingly endless list of other projects. But it sure was fun while it lasted.

EARTH'S OLDEST "FOSSILS"?

What does a negative carbon-isotope value tell us about an ancient rock? This question came into focus following a surprising announcement in the November 7, 1996, issue of *Nature* of the discovery of Earth's most ancient fossil carbon. The Earth's oldest known rocks, 3.85-billion-year-old banded-iron outcrops from the remote island of Akilia off the southwest coast of Greenland, reveal not the slightest trace of anything that looks like a fossil. Nevertheless, these rocks may contain a modest store of carbon. Even though the rocks have experienced severe alteration through the ravages of temperature, pressure, and time, some of that carbon is encased in the protective mineral apatite. When Scripps Institution of Oceanography geochemist Stephen J. Mojzsis (now at the University of Colorado) and his colleagues collected those rocks and performed the first carbon-isotope analysis at UCLA in 1996, they were delighted to find light carbon, on average a dramatic 3.7 percent lighter than reference limestone. No known abiotic process produces that kind of value. That simple number, −37, was enough to convince many geologists that life had achieved a firm foothold by that ancient date.

Such a result did much more than establish a world record for ancient life. The work of Mojzsis and his colleagues seemed to narrow the window for life's origin, which presumably couldn't have emerged until after the last global sterilizing asteroid impact, roughly 4 billion years ago. If signs of life persist in 3.85-billion-year-old rocks, then life arose very quickly indeed.

But, as it turned out, the Akilia rocks posed problems. Earth's oldest rocks have been through a lot: heated and squeezed and contorted beyond belief. Billions of years inevitably alter the fabric of a rock. Mojzsis interpreted the Akilia formation, with its appearance of intensely folded layers, as metamorphosed oceanic sediments—a perfectly reasonable residence for early cellular life. But when geologists

Christopher Fedo of George Washington University and Martin Whitehouse of the Swedish Museum of Natural History performed a more detailed geological analysis of the carbon-bearing outcrop, the rocks proved to be an ancient molten igneous mass that gradually solidified deep underground from temperatures approaching 1,000°C. The carbon deposits must have formed under extreme metamorphic conditions deep in the crust. Under no circumstance could those rocks have contained life at the time of their formation.

Scientists quickly came up with a range of plausible explanations for the light carbon. Heating experiments, which preferentially release recent organic contaminants, revealed that some of the rocks' carbon is modern. It's also possible that some natural *non*biological processes also generate light carbon. Today much of the carbon cycle is regulated by life, and all carbon compounds derived from living organisms are isotopically light. But before the first microbial life, there could have been equally vigorous geochemical processes that separated carbon-12 from carbon-13. If so, then isotopes alone can provide scant help in recognizing life in Earth's most ancient formations—or from rocks on other worlds, for that matter.

At best, the isotopic evidence from Greenland is ambiguous. And so, in their search for unambiguous proof of ancient life, paleontologists have had to turn to even more elusive fossils—fragments of life's oldest biomolecules.

5

Idiosyncrasies

The ability of the major atomic components of the cell to combine into molecules of considerable complexity . . . is enormous. However, the actual number of compounds that are used in biology is relatively small, comprising only hundreds of compounds.

Noam Lahav, *Biogenesis*, 1999

In a sense, you are what you eat. Nutrition labels on the side of every packaged food underscore this biochemical fact: Fats, carbohydrates, and proteins satisfy life's energy requirements (i.e., calories) and provide life's most basic molecular building blocks as well.

Carbon, the essential element of life, combines with other atoms in every living cell to form the molecules of life. Even as ancient rocks can entomb the original carbon isotopes from cells, so too, under the right circumstances, they can preserve larger fragments of life's biomolecules. Such molecular remnants hold great promise for identifying ancient life, because terrestrial life is so remarkably, uniquely idiosyncratic in its choice of chemical building blocks.

SYNTHETIC QUIRKINESS

Consider the example of life's hydrocarbons, the molecular family that includes waxes, soaps, oils, and all manner of fuels, from gasoline to Sterno. All cells require a rich variety of these molecules, which incorporate long, chainlike segments of carbon and hydrogen atoms. Hydrocarbons, which we eat in the form of fats and oils, serve many cellular functions, including the production of flexible cell membranes, efficient energy storage, varied internal support structures, and more.

In life and in commerce, long hydrocarbon molecules are usually made by linking smaller pieces end-to-end. When industrial chemists want to synthesize hydrocarbons, or when these molecules arise by natural nonbiological processes, the molecular chains are usually lengthened one carbon group at a time. This process ordinarily yields a suite of molecules with many different lengths, from just a few to many dozens of carbon atoms long, but all formed by the same stepwise mechanism.

Life builds hydrocarbons differently and in a strikingly idiosyncratic way. In each cell an amazing tool kit featuring half-a-dozen different protein catalysts, collectively called the "fatty acid synthase," facilitates the assembly of hydrocarbon chains by adding units of three carbon atoms to a growing chain and then stripping one away. The net result is carbon addition by pairs. So life's biochemistry is often characterized by a preponderance of hydrocarbon chains with an even number of carbon atoms: chains of 12, 14, or 16 carbon atoms occur in preference to 11, 13, or 15. As a result, given a suite of molecules from some unknown source and a mass spectrometer that can analyze the size distribution of those molecules, it's not too difficult to tell whether the hydrocarbons came from living cells or from nonbiological processes.

Polycyclic compounds, an even more dramatic example of life's molecular idiosyncrasies, include a diverse group of carbon-based molecules with several interlocking 5- and 6-member rings. A variety of cyclic molecules are found everywhere in our environment. Even before Earth was born, they were produced abundantly by chemical reactions in interstellar space and during star formation—processes that littered the cosmos and seeded the primitive Earth with cyclic organic molecules. The PAHs (polycyclic aromatic hydrocarbons) found in the Martian meteorite ALH84001 are examples of these ubiquitous compounds. Cyclic molecules continue to be synthesized on Earth as an inescapable by-product of all sorts of burning: They are found in the soot of fireplaces and candles, the smoke of incinerators and forest fires, and the exhaust of diesel engines. Travel to the remotest places on Earth—the driest deserts of North Africa, deep ocean sediments, even Antarctic ice—and you'll find PAHs.

Every living cell manufactures a variety of polycyclic carbon compounds but, as with hydrocarbon chains, the polycyclic compounds produced by life are much less varied than those produced by inorganic processes. The 4-ring molecules called sterols, including cholesterol, steroids, and a host of other vital biomolecules, underscore this point. Literally hundreds of different 4-ring molecules are possible, yet while the relatively random processes of combustion or interstellar synthesis yield a complex mixture of cyclic compounds, life zeroes in almost exclusively on sterols and their by-products.

Again, cells employ a remarkably quirky synthesis pathway. The first step in forming a sterol is to manufacture lots of isoprene, a 5-carbon branching molecule (which the cell makes from three smaller molecules). Six isoprene molecules line up end-to-end to form

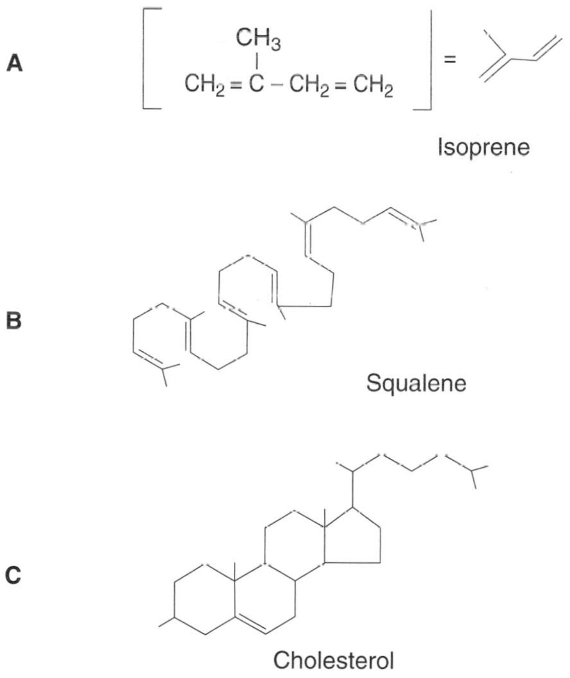

A
$$\left[\begin{array}{c} CH_3 \\ | \\ CH_2 = C - CH_2 = CH_2 \end{array} \right] = $$

Isoprene

B

Squalene

C

Cholesterol

Cells manufacture polycyclic molecules in an idiosyncratic three-step process. First, three small molecules link together to form isoprene (A). Then six isoprene molecules line up end-to-end to make squalene (B). Finally, squalene folds up into the 4-ring cholesterol molecule (C). In these and subsequent drawings of molecules, each short line segment represents a chemical bond between two carbon atoms.

squalene, with 30 carbon atoms—24 of them in a chain, with six single carbon atoms branching off at regular intervals. This long molecule then folds up into the 4-ring sterol backbone.

Biochemical textbooks describe dozens of other examples of elaborate synthetic pathways: photosynthesis to make the sugar glucose, glycolysis (splitting glucose) to make the energy-rich molecule ATP (adenosine triphosphate), metabolism via the citric acid cycle, the production of urea, and countless other vital chemical processes. Over and over, we find that cells zero in on a few key molecules. DNA and RNA, which carry the genetic code, rely on ribose and deoxyribose alone, eschewing the dozens of other 5-carbon sugars. Proteins are constructed from only 20 of the hundreds of known amino acids. What's more, sugars and amino acids often come in mirror-image pairs—so-called "right-handed" and "left-handed" variants—but life uses right-handed sugars and left-handed amino acids almost exclusively.

The take-home lesson is that life is exceedingly choosy about its chemistry. Of the millions of known organic molecules with up to a dozen carbon atoms, cells typically employ just a few hundred. This selectivity is perhaps the single most diagnostic characteristic of living versus nonliving systems. If an ancient rock is found to hold a diverse and nondescript suite of organic molecules, then there's little we can conclude, yea or nay, about its biological origins. It may once have held life, or it may simply represent an abiotic accumulation of organic junk. If, on the other hand, an old rock holds a highly selective suite of carbon-based molecules—predominantly even-numbered hydrocarbon chains or left-handed amino acids, for example—then that's strong evidence that life was involved.

A crucial requirement, if this logic is to be implemented in the search for life here or on other worlds, is that biomolecules must be stable over time spans of billions of years. Large protein molecules won't last that long, and neither will the 20 amino acids that comprise the building blocks of proteins. Nor will most carbohydrates or hydrocarbon chains. Over time, water attacks the bonds of these biomolecules, breaking them into smaller fragments of no diagnostic use. But polycyclic compounds, like sterols, degrade more slowly and might survive over geological spans of time. Therein lies a possible top-down path to the discovery of life that is distant in space or time.

THE HOPANE STORY

Once in a very great while, extremely old rocks are found to hold microscopic droplets of a petroleum-like black residue—hydrocarbons that represent the remains of ancient marine algae. When such droplets were first discovered, decades ago, most scientists discounted the possibility that these organic remains were very old; no oil could survive billions of years of geological processing, they said. But subsequent discoveries and improved analytical techniques have convinced the geological community that a hardy breed of organic hydrocarbons can survive in ancient rock provided that temperatures never got too high.

In their quest for life signs, a group of Australian scientists has focused upon what are perhaps the ideal biomarkers—distinctive sterol-derived polycyclic hydrocarbon molecules called hopanes. This group of elegant 5-ring molecules is known in nature only from the biochemical processes of cellular life, where it concentrates in protective cell membranes. Furthermore, different variants of hopanes point to specific groups of microbes with distinctive biochemical lifestyles. If an ancient rock happens to encase and preserve hopane-related molecular fragments with the diagnostic structures of once-living biomolecules, then we have convincing evidence of ancient life.

In 1999, a team of scientists led by Roger Summons (then at the Australian Geological Survey Organisation) presented compelling evidence for the survival of hopanes in a sequence of 2.7-billion-year-old

This 5-ring structure is characteristic of hopane, a distinctive biomolecule whose backbone may be preserved for billions of years in ancient sediments.

sedimentary rocks called the Pilbara Craton, in Western Australia. The black, carbon-rich shale layers in question came from a section of drill core extracted from a depth of about 700 meters. The mineralogy of the shale revealed that it had never experienced a temperature higher than about 300°C—an unusually benign history for such an ancient deposit.

Hopanes have been common biomolecules for a long time, so the Australian team's principal challenge was ruling out contamination from more recent life. The rocks might have been contaminated hundreds of millions of years ago by subsurface microbes, or by groundwater carrying biomolecules from the surface, or perhaps even by oil, seeping from some other sedimentary horizon. Summons and his colleagues discounted the last of these possibilities because they found no trace of petroleum in adjacent sediment layers. Modern contamination from living cells, which abound in the lubricants that scientists use to drill their deep holes in the host rock, was also a concern. The team ruled out such contamination, too, because the suite of molecules preserved in the shale was "mature," containing none of the fragile organic species that would point to recent lubricants and accompanying microbial activity.

Summons and his co-workers had to develop meticulous procedures to expose and clean unadulterated fresh rock surfaces: Break the rock, wash the surface, and measure the wash for contamination. They resorted to smaller and smaller rock fragments to avoid the inevitable impurities that had seeped in along cracks. Summons found that properly prepared powdered shale contained a distinct suite of ancient hydrocarbon molecules at hundreds of times higher concentrations than in adjacent chert and basalt layers from the same drill core. Nevertheless, the amount of hopanes was minuscule: of all the carbon-rich material extracted from the rock, no more than a precious few hundred parts per million were hopanes and related polycyclic molecules. Still, the very presence of hopanes provided evidence for ancient microbial life.

Having overcome daunting hurdles, Summons and his colleagues announced their finding in August 1999, in two remarkable papers, one in *Science* and the other in *Nature*. The *Science* article detailed extraction of hopanes from 2.7-billion-year-old shale—results that broke the previous record for the oldest molecular biomarker by about a billion years. The *Nature* article described the discovery of hopanes from

2.5-billion-year-old Australian black shale—a younger nearby formation, but with a twist. That formation included 2-methylhopanoid, a
hopane variant known to occur in cyanobacteria, which are the primitive photosynthetic microbes responsible for generating Earth's oxygen-rich atmosphere. The Australian team had found suggestive
evidence that cyanobacteria were thriving long before 2 billion years
ago, when Earth's atmosphere is thought to have achieved modern levels of oxygen.

By extracting and identifying unambiguous biomarkers in ancient
rocks, Summons and colleagues had made a major advance in detecting and characterizing ancient life. They also helped close the gap in
our ignorance of life's emergence by embellishing the top-down story
and pushing it just a little bit further back in time.

BIOSIGNATURES AND ABIOSIGNATURES

The quest for unambiguous "biosignatures," including hopanes and
other distinctive molecules, represents an effective strategy in the
search for ancient life on Earth and other worlds. However, the identification of "abiosignatures" —chemical evidence that life was *never*
present in a particular environment—might also prove important in
constraining models of life's emergence.

Abiosignatures hold special significance to astrobiologists, who
search for life in Martian meteorites and other exotic specimens. Are
there physical or chemical tests that might *preclude* the presence of
past life in those specimens? "NO LIFE ON MARS!" would be a bum-
mer of a headline, but would nevertheless carry great scientific, not to
mention philosophical, implications about the frequency of life's
emergence.

Hopanes not only represent biosignatures for ancient life on Earth,
but they also point to a search strategy for other worlds, especially our
nearest neighbor, Mars. Based on what we now know about life and its
fossil preservation, we are unlikely to find unambiguous Martian fossils of single cells, much less animals or plants, at least not any time
soon. A Mars sample return mission won't happen for at least a dozen
years, while human exploration of the red planet is many decades away.
And even with such hands-on exploration, we'd be incredibly lucky to
find a convincing fossil. We're much more likely to find local concentrations of carbon-based molecules, from which we can determine the

carbon-isotopic composition. However, as the Greenland incident reveals, a simple isotopic ratio may not be sufficient to distinguish nonbiological chemical systems from those that were once living.

Suites of carbon-based molecules, if we can find them, hold much greater promise. An array of molecular fragments derived from a colony of cells, if not too degraded, will differ fundamentally from a geochemical suite synthesized in the absence of life. For now, molecules represent our best hope of finding proof of life both here and elsewhere in our solar system.

The ideal molecular biosignatures—and abiosignatures as well—must display three key characteristics. First, biosignatures should consist of distinctive molecules or their diagnostic fragments that are essential to cellular processes. Similarly, abiosignatures should consist of molecules that clearly point to nonbiological processes.

The second criterion is stability: biosignatures—and abiosignatures as well—must be molecules able to survive through geological time. Even the least altered ancient sediments have been subjected to billions of years of temperatures greater than the boiling point of water—conditions that significantly alter the chemical characteristics of any suite of organic molecules, whether biological or not. This criterion of stability, consequently, focuses our attention on unusually stable molecules.

Finally, the molecules must occur commonly and in reasonable abundance. A molecular biosignature or abiosignature is of no use unless it can be detected by mass spectrometry or other standard analytical techniques.

Hopanes, and the related sterols, are unquestionably excellent diagnostic biosignatures from the standpoint of stability, and they're reasonably easy to analyze. Many ancient deposits yield traces of these molecules, and they will continue to be a tempting target for analysis, as well as a model for finding other biosignatures. But hopanes are probably not the ultimate answer in the search for signs of life: They seldom occur in abundance, and their absence cannot be taken as a reliable abiosignature.

An alternative to the search for reliable biosignatures and abiosignatures might be to identify diagnostic ratios of molecular fragments, akin to the carbon-12/carbon-13 isotopic ratio. However, we're confronted with a vast multitude of possible molecule pairs. Which pair of molecules should we study?

My first foray into the search for biomarkers occurred in the summer of 2004. Preliminary studies by George Cody on organic compounds in meteorites prompted us to look at the ratio of two of the commonest PAHs: anthracene and phenanthrene. These 3-ring polycyclic molecules, both made up of 14 carbon atoms and 10 hydrogen atoms ($C_{14}H_{10}$), differ only in the arrangement of the rings: In anthracene the rings form a line, in phenanthrene a dogleg. We realized that the ratio of these two molecules might fulfill the essential biomarker requirements: Both are distinctive, relatively stable, common in the geological record, and easy to detect in trace amounts.

Phenanthrene and anthracene form in abundance through a variety of nonbiological processes, including any burning process that produces soot. These cyclic compounds are also synthesized in deep space, where they contribute to the molecular inventory of the carbon-rich meteorites called carbonaceous chondrites. The most celebrated of these is the Murchison meteorite, which fell to Earth in a cow field outside the small town of Murchison, about 100 miles north of Melbourne, Australia, on September 28, 1969. Meteorites hit Earth all the time, but the Murchison fall was special. For one thing, it was big—several kilograms of rock. For another, it was fresh and relatively un-

Phenathrene (top) and anthracene are 3-ring polycyclic molecules that differ only in their shape. The ratio of these two molecules differs in abiotic and biological systems.

contaminated—a number of pieces were collected while they were still warm. But, most important, the Murchison was a carbonaceous chondrite, containing more than 3 percent by weight of organic molecules. That black, resinous matter, formed billions of years ago in dense molecular clouds and protoplanetary disks, held a treasure trove of the molecules that could have accumulated on the prebiotic Earth.

George Cody had found that such meteorites often display about a 1:1 ratio of phenanthrene to anthracene. But biochemical processes seem to produce a different ratio. Many polycyclic biomolecules—including sterols and the varied hopanes—incorporate a 3-ring dogleg, so phenanthrene is a common and expected biomolecular fragment, and it should persist in rocks and soils, even when larger molecules break down. But for some reason, life almost never uses anthracene's linear arrangement of three rings. Anthracene would thus seem to be correspondingly rare as a biomolecular fragment. Cody had found that biogenic coals typically hold 10 times more phenanthrene than anthracene.

Is the ratio of phenanthrene to anthracene a useful biomarker? Testing this idea required measuring the ratios of cyclic compounds in lots of samples, so that was the task I gave to Rachel Dunham, a bright and energetic undergraduate summer intern from Amherst College. Over the course of her 10-week stay in Washington, Rachel assembled dozens of natural and synthetic PAH-containing samples from around the world, analyzed them with our gas chromatograph/mass spectrometer, and managed to track down many more analyses from the vast coal and petroleum literature, since it turns out that PAHs are especially abundant in some fossil fuels.

The first few data points seemed to support the hypothesis. The Murchison and Allan Hills meteorites showed phenanthrene-to-anthracene ratios of 1.7:1 and 2:1, respectively. The biogenic Burgess Shale and a mature coal, on the other hand, yielded much higher ratios, close to 15:1. But then results began to scatter. Some low-grade coals had ratios less than 5:1, while the black, fossil-rich Enspel Shale was only about 2:1. The promising hypothesis began to crumble.

After several weeks of effort, and a thorough review of the published literature, Rachel discovered that we were simply reinventing the wheel. Coal experts have long known that anthracene is slightly less stable than phenanthrene. Consequently, "high-grade" coals that have experienced prolonged high temperatures and pressures have a higher ratio of phenanthrene to anthracene, topping 20:1 in some specimens. Such a high ratio in any sample may have resulted from prolonged heating and have nothing at all to do with a biogenic past.

Our only useful conclusion was that unambiguously biological specimens *never* seem to display ratios less than about 2. So a lower ratio of phenanthrene to anthracene, as found in meteorites, synthetic-run products, and soot from burning carbon, may provide a valid abiomarker.

The next step? Perhaps we'll try another PAH ratio, such as that of phenanthrene to pyrene, a particularly stable diamond-shaped molecule with four interlocking rings ($C_{16}H_{10}$). One thing is certain: we'll never run out of molecules to try.

LOOKING FOR LIFE ON MARS

The most exciting, important, and potentially accessible field area to look for ancient alien life is Mars, which, like Earth, formed some 4.5 billion years ago. A mass of new data points to an abundance of surface water during the planet's first billion years, dubbed the Noachian epoch by Mars geologists. Sunlit lakes, hydrothermal volcanic systems, and a benign temperature and atmosphere might have sparked life and made Mars habitable long before Earth. Perhaps fossils, molecular and otherwise, litter the surface.

The quest for Martian life has a checkered history. A century ago, American astronomer Percival Lowell reported observations of a network of canals on the red planet—evidence of an advanced civilization, he thought. Such speculation fueled the imaginations of science fiction writers, but hardened the scientific community to such unsubstantiated claims. NASA's remarkable Viking Mars lander of the mid-1970s carried an array of experiments designed to find organic compounds and to detect cellular activity, but ambiguous results merely led to more controversy. Hot debates over purported fossils in the Allan Hills Martian meteorite represent just one more chapter in this contentious saga. Given such a troubled context, NASA will choose its next round of life detection experiments with the greatest of care.

Humans aren't going to set foot on Mars anytime soon, but that's what NASA's amazing rovers are for. *Sojourner, Spirit, Opportunity,* and other robotic vehicles provide the experimental platform; they haul the instruments across the desolate Martian surface to probe tantalizing rocks and soils. The key to finding life (or the lack thereof) is to design and build a flight-worthy chemical analyzer for life. That has been the occupation of Andrew Steele throughout much of his career.

"Steelie," my ebullient colleague at the Geophysical Laboratory, is a microbial ecologist who got his start studying the microbial corrosion of stainless steel in nuclear reactors. British Nuclear Fuels Ltd. sponsored his PhD thesis, which is still largely classified and unpublished. Radioactive isotopes often contaminate the thin outer layer of stainless steel that lines nuclear reactors—a difficult and costly clean-up problem. Steelie invented new microscopic techniques to study the steel surfaces, and he found that some microbes secrete biofilms that rapidly eat away steel, thus stripping off the affected layers and greatly simplifying decontamination. [Plate 4]

In 1996, just two weeks after defending his PhD thesis in England, the Allan Hills meteorite story broke. The timing was perfect, and Steelie was hooked. Setting his sights on nonradioactive ecosystems, he became obsessed with attempts to detect ancient life from the faint molecular traces in rocks. He contacted the NASA team, who were lacking in microbiology expertise and thus eager for his help. David McKay sent him a sample of the precious Martian rock, and within a year Steele had applied his microscopic techniques to investigations of the purported microbes. He made a memorable presentation at NASA's annual Lunar and Planetary Science Conference in Houston and landed himself a job as a NASA microbiologist working at the Johnson Space Center under McKay's guidance.

Steelie's assignment at NASA was to head the JSC Blue Team (consisting of Steele and a fellow gadfly), who were to try out every possible idea to disprove the hypothesis of Martian life in the Allan Hills meteorite. McKay led the significantly larger competing Red Team. McKay's strategy of examining both sides of the issue was laudable and in the best tradition of scientific objectivity, but it may have backfired when Steelie did his job too well. He performed a series of high-resolution microscope studies and by 1998 had discovered that the Allan Hills specimen (like virtually every other meteorite he had ever examined) was riddled with contaminating Earth microbes. McKay was unconvinced and continued to promote his original interpretation of ALH84001. Discouraged at the cool reception accorded his findings, and missing his wife and family in England, Steele left his steady NASA employment in early 1999 for a hectic schedule of visiting professorships and research jobs at the Universities of Montana, Oxford, and Portsmouth, along with more NASA consulting.

About the time that Andrew Steele's research was making him something of a *persona non grata* at the Johnson Space Center, the Carnegie Institution's Geophysical Laboratory found itself under new leadership. Wes Huntress, the former associate administrator of the NASA Office of Space Science (and the man who introduced the term "astrobiology" to the NASA community), had been hired to shake things up and expand Carnegie's fledgling astrobiology effort. In a bold and welcome move, Huntress made Steele his first staff appointment in 2001. Steelie's peripatetic scientific lifestyle seemed unsuited to the traditional academic world, but his unconventional background was just the ticket for the Geophysical Lab.

He arrived like a whirlwind, his shoulder-length blond hair and open tie-dyed lab coat flying behind him as he dashed between office and lab. Crates and boxes arrived by the dozen at his second-floor domain, two doors down from my own office. He crammed his lab space with DNA sequencers, chip writers and chip readers (benchtop machines that prepare and read slides), and a bewildering array of other microbiological hardware, most of which none of us geologists had ever seen before. A small army of bustling postdocs followed and the corridor took on new life. Chemist Mark Friese brought his collection of orchids to grace one alcove with a forest of exotic blooms. Molecular biologist Jake Maule maintained his pro-level golf game by challenging all comers to putting contests—a regular stream of golf balls began rolling past my open doorway.

In such an environment, new ideas fly thick and fast. Steelie had a master plan. He wanted to build a life-detecting machine to fly to Mars—a huge interdisciplinary project, given how much we still don't know. On the scientific side, he had to figure out what constitutes a legitimate biosignature, so Steelie embarked on various paleontology projects, principally with German postdoc Jan Toporski, who happens to be his brother-in-law and soccer buddy. (There are also a lot of soccer balls and other soccer paraphernalia in the corridor.) They focused on a 25-million-year-old fossil lake in Enspel, Germany, where well-preserved fish, tadpoles, and other animals are found. Studies of their molecular preservation would provide important hints about life's most stable and diagnostic molecular markers.

Then there was the detection part. It was essential to develop an unambiguous procedure to find the tiniest amounts of any target molecule. That's where Jake Maule came in. An aspiring astronaut trained in clinical medical technology, Jake's job was to develop molecular antibodies—proteins with specialized shapes that would lock onto only one type of target molecule. [Plate 4]

Jake focused on producing hopane antibodies for use in a rapid and sensitive field test for microbes. His procedure involved injecting mice with hopanes and letting their immune systems do most of the work. Hopanes are too small to evoke an immune response by themselves, so Jake attached hopane molecules to a big protein called BSA (for bovine serum albumin). He injected 30 mice with a hopane–BSA solution, waited a week, and gave each a booster shot. In about three weeks, the mice manufactured a suite of hopane-sensitive antibodies. From each mouse, Jake extracted a syringe full of blood, which was centrifuged to separate out the red and white blood cells from the watery fluid that held the antibodies. Ultimately, each mouse yielded one tiny, precious droplet of that antibody-rich fluid.

Armed with hopane antibodies, Maule was ready to analyze ancient rocks, in an elegant four-step process.

Step 1: He crushed various promising rock samples and washed them in a solvent, which concentrated any hopane residues. Those solutions were loaded into the chip writer, a sleek benchtop machine about the size of a breadbox that placed an array of tiny dots of the solutions (some containing hopane and others not) and various standards onto a glass slide.

Step 2: With the chip writer, he applied a tiny amount of hopane antibody onto each of the sample spots, then rinsed. Some spots then retained hopanes with attached antibodies, while other spots were washed clean.

Step 3: He then treated each spot with another solution, this one containing a second antibody that locks onto any mouse antibody and is also highly fluorescent. All the spots with hopanes attached to mouse antibodies would thus fluoresce.

Step 4: Finally, he took the glass slide with its array of spots and put it into a second sleek machine, the chip reader, which, as its name suggests, recorded which spots fluoresced and which spots didn't. Like microscopic lightbulbs, his antibodies glowed when hopanes were present in the rock sample.

Jake had devised a fast, automated process for identifying hopanes. But all that work and more was mere prelude to the most crucial aspect of a flight-ready analytical instrument—the design and engineering. Concepts and benchtop demonstrations were one thing, but an instrument on Mars has to be absolutely reliable, shock resistant, and very, very small. Steelie began dealing with nitty-gritty questions of how to collect a soil sample, how to introduce a small amount of sterile solvent to dissolve the target molecules, how to excite a fluorescent signal, and how to relay the information home to Earth, all in a tiny box. Soon, armed with a million dollars of NASA funding, he planned to fly the instrument on NASA's 2013 mission to Mars.

Steelie's baby is called MASSE—the Microarray Assay for Solar System Exploration. Adapting the latest in chip writer/chip reader technology, his team is building both flight-ready and handheld devices that use antibodies to detect trace amounts of dozens of diagnostic biomolecules: hopanes, sterols, DNA, amino acids, a variety of proteins, even rocket exhaust. Carnegie is not alone in developing such an instrument, and given the intense competition of other dedicated design teams, there's no guarantee that MASSE will ever fly. Steelie is also competing against a seductive sample-return mission—a technically challenging effort to return a soda can-sized canister of Martian rock and soil to Earth on the same 2013 mission. Only one instrument package will be selected. Even if MASSE does fly, there's no guarantee that it will arrive safely, and if it does arrive safely that it will find anything of interest. Of course, Steelie and his group are learning lots of fascinating stuff along the way. And I've never seen scientists have so much fun.

In our quest to understand life's emergence, fossils provide essential clues. Even lacking morphological evidence, fossil elements, isotopes, and molecules point to the nature of primitive biochemical processes. These microscopic remains also reveal a diversity of life-supporting environments and help to constrain the timing of life's genesis. What's more, if we're lucky, fossils from Mars or some other extraterrestrial body may eventually provide the best evidence that life has emerged more than once in the universe.

And yet, valuable as these insights may be, all known fossils represent remains of advanced cellular organisms similar to those alive today. Few, if any, clues remain regarding the emergent biochemical steps that must have preceded cells. To understand how life arose, therefore, we must go back to the beginning and approach the question of origins from the bottom up.

Interlude—God in the Gaps

Darwinists rarely mention the whale because it presents them with one of their most insoluble problems. They believe that somehow a whale must have evolved from an ordinary land-dwelling animal, which took to the sea and lost its legs.
. . . . A land mammal that was in the process of becoming a whale would fall between two stools—it would not be fitted for life on land or sea, and would have no hope of survival.

Alan Haywood, 1985

This rant by Alan Haywood, and similar silly statements by his creationist colleagues, reveals a deep mistrust of the scientific description of life's emergence and evolution. But he is correct in one sense. Science is a slave to rigorous logic and inexorable continuity of argument. If life has changed over time, evolving from a single common ancestor to today's biological diversity, then many specific predictions about intermediate life-forms must follow. For example, transitional forms between land mammals and whales must have existed sometime in the past. According to the creationists in the mid-1980s, the lack of such distinctive forms stood as an embarrassing, indeed glaring, proof of evolution's failure. Their conclusion: God, not Darwinian evolution, must have bridged the gap between land and marine animals. But what appeared to them as an embarrassment for science then, has since underscored the power of the scientific method.

Science differs from other ways of knowing because scientific reasoning leads to unambiguous, *testable* predictions. As Haywood so presciently predicted, whales with atrophied hind legs must have once swum in the seas. If Darwin is correct, then somewhere their fossils must lie buried. Furthermore, those strange creatures must have arisen during a relatively narrow interval of geological time, bounded by the era before the earliest known marine mammals (about 60 million years ago) and the appearance of streamlined whales of the present era (which appear in the fossil record during the past 30 million years). Armed with these predictions, several paleontologists plotted expeditions into the field and targeted their search on shallow marine formations from the crucial gap between 35 and 55 million years ago for new evidence in the fossil record. Sure enough, in the past decade paleontologists have excavated more than a dozen of these "missing links" in the development of the whale—curious creatures that sport combinations of anatomical features characteristic to both land and sea mammals.

Moving back in time, one such intermediate form is the 35-million-year-old *Basilosaurus*—a sleek, powerful, toothed whale. This creature has been known for more than a century, but a recent discovery of an unusually complete specimen in Egypt for the first time included tiny, delicate vestigial hind leg bones. That's a feature without obvious function in the whale, but such atrophied legs provide a direct link to four-limbed ancestral land mammals.

And then a more primitive whale, *Rodhocetus*, discovered in 1994 in Pakistani sediments about 46 million years old, has more exaggerated hind legs, not unlike those of a seal. And in that same year paleontologists reported the new genus *Ambulocetus*, the "walking whale." This awkwardly beautiful 52-million-year-old creature represents a true intermediate between land and sea mammals.

Nor does the story end there. In September 2001, the cover stories of both prestigious weekly magazines *Science* and *Nature* trumpeted the discovery of a new proto-whale species that had just been reported from rocks about 50 million years old. *Nature's* cover story, titled "When whales walked the earth," underscores the power of science and the futility of the creationists' task. Science makes specific, testable predictions. Anyone can go out into the natural world and test those predictions. The creationists were wrong.

Today's creationists have toted out a new version of this old

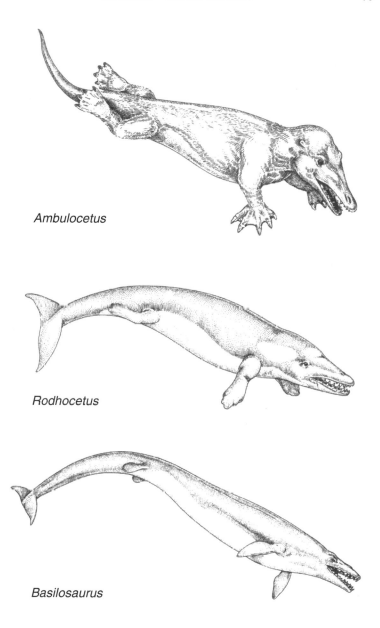

Ambulocetus

Rodhocetus

Basilosaurus

The evolution of whales is illustrated by recent fossil finds, including *Ambulocetus* (52 million years old), *Rodhocetus* (46 million years old), and *Basilosaurus* (35 million years old) (from National Academy of Sciences, 1999).

"God-in-the-gaps" argument under the fancy name "intelligent design." Their argument goes like this. Life is so incredibly complex and intricate that it must have been engineered by a higher being. No random natural process could possibly lead from non-life to even the simplest cell, much less humans. The promoters, notably Michael Behe and William Dembski, don't talk about "God," but they leave open the question of who designed the designers.

Such an argument is fatally flawed. For one thing, intelligent design ignores the power of emergence to transform natural systems without conscious intervention. We observe emergent complexity arising all around us, all the time. True, we don't yet know all the details of life's genesis story, but why resort to an unknowable alien intelligence when natural laws appear to be sufficient?

I also see a deeper problem with intelligent design, which I believe trivializes God. Why do we have to invoke God every time we don't have a complete scientific explanation? I am unpersuaded by a God who must be called upon to fill in the gaps of our ignorance—between a cow and a whale, for example. The problem with this view is that as we learn more, the gaps narrow. As paleontologists continue to unearth new intermediate transitional forms, God's role is squeezed down to ever more trivial variations and inconsequential modifications.

Isn't it more satisfying to believe in a God who created the whole shebang from the outset—a God of natural laws who stepped back and doesn't meddle in our affairs? In the beginning God set the entire magnificent fabric of the universe into motion. Atoms and stars and cells and consciousness emerged inexorably, as did the intellect to discover laws of nature through a natural process of self-awareness and discovery. In such a universe, scientific study provides a glimpse of creator as well as creation.

Part II

The Emergence of Biomolecules

The top-down study of life's origin—via the examination of ancient fossils—doesn't tell us much about early biochemistry. Fossil cells, molecules, and isotopes are indistinguishable from those of contemporary life-forms; consequently, from them we can gain detailed knowledge only of the final state—modern cellular life.

Nevertheless, common sense points to the essential first step in life's emergence—the synthesis and accumulation of abundant organic (that is, carbon-based) molecules, such as amino acids, sugars, and other vital molecules of which life would eventually be made. In the beginning, life's raw materials consisted of water, rock, and the simplest and most basic volcanic gases—carbon dioxide, nitrogen, and maybe a little methane and ammonia. Add energy to the mix and organic molecules emerge.

The experimental pursuit of this ancient process, arguably the best understood aspect of life's chemical origins, began in earnest just over a half-century ago, with the pioneering studies of University of Chicago graduate student Stanley Miller and his distinguished mentor, Harold Urey. Together they demonstrated the organic synthesis that occurred as Earth's primitive atmosphere and ocean were subjected to bolts of lightning and the Sun's intense radiation. Chemical processes in deep space, in the solar nebula, during asteroid and comet impacts, and even within the pressure cooker of Earth's deep interior, also generated abundant organic molecules. By 4 billion years ago, Earth's globe-spanning ocean must have become a complex, albeit dilute, soup of life's building blocks. Though not alive, this chemical system was poised to undergo a sequence of increasingly complex stages of molecular organization and evolution.

6

Stanley Miller's Spark of Genius

The idea that the organic compounds that serve as the basis of life were formed when the earth had an atmosphere of methane, ammonia, water, and hydrogen instead of carbon dioxide, oxygen, and water was suggested by Oparin and has been given emphasis recently by Urey and Bernal. In order to test this hypothesis an apparatus was built. . . .
                    Stanley L. Miller, *Science*, May 15, 1953

The experimental investigation of life's origin is a surprisingly recent game. Two centuries ago, many scientists accepted the intuitively reasonable idea, championed by Aristotle and a pantheon of other ancient scholars, that a life force permeates the cosmos. A central precept of this doctrine, known as vitalism, is that life arises spontaneously all around us, all the time. The question of life's origin wasn't asked—at least not in the modern experimental sense.

Think about your own experience and you'll see why this idea of spontaneous generation isn't such a strange notion. Mold seems to grow spontaneously on bread, maggots appear as if by magic in old meat, and every spring new plants sprout and grow in the season of renewal. It's not surprising that so many people, for so long, accepted unquestioningly the spontaneous generation of life.

This view continued well into the nineteenth century, though not without a number of respected dissenters and considerable debate. The seventeenth-century invention of the microscope and the subsequent realization that microscopic life abounds complicated the story, but failed to resolve the controversy. Those who favored spontaneous generation saw microbes as just another manifestation of the life force;

those opposed saw microscopic life as an obvious source of experimental contamination.

The interpretation of experiments is rarely unambiguous, as highlighted by a delicious eighteenth-century exchange. The noted Italian physiologist Lazzaro Spallanzani explored spontaneous generation by comparing the behavior of boiled and unboiled sealed flasks, each filled with bacteria-infested, nutrient-rich water. He found that flasks boiled for an hour remained sterile indefinitely as long as they remained sealed, whereas unboiled flasks quickly became cloudy with microbes. Spallanzani concluded that ubiquitous microscopic life-forms contaminate any unsterilized experiment.

Englishman John Needham, an astute amateur experimenter who conducted his own fragrant experiments on hot mutton gravy, came to a different conclusion. Needham agreed that boiling kills microbes, but microbes soon reappeared in abundance when his gravy cooled. He argued that these cells arose by spontaneous generation. Spallanzani countered that Needham's microbes came from airborne contamination and, to prove his point, he undertook a new series of experiments in which he demonstrated that no microbes appeared when he pumped air out of his flasks and then boiled the water. Needham disagreed: A property of the air, not the water, must carry the life force, he said.

We react to this historical incident with a biased worldview. Of course Spallanzani's conclusions about microbial contamination were right and Needham's support of spontaneous generation was wrong, we are tempted to say. But a naïve and impartial observer of the time, faced with these conflicting claims, would have had a difficult time choosing between invisible microbes and an invisible life force. Both arguments were internally consistent, so doubts remained.

The influential French chemist Louis Pasteur helped to abolish vitalism and the theory of spontaneous generation once and for all with his own brilliant series of experiments on sterilization. He prepared a nutrient-rich sugar solution and poured it into several beakers. One set of beakers was tightly sealed to prevent any contact with the ambient air. Other beakers were left open with a narrow twisting neck, so that the sugar solution was in contact with the ambient air, but unlikely to be contaminated by stray microbes, which were unable to traverse the long glass passage. As a control, he left other beakers wide open or deliberately contaminated them with ordinary dust. He showed that boiled water, if isolated from airborne microbes, remained sterile indefinitely. Only microbial contamination causes new growth, not spontaneous generation—a result of tremendous practical as well as theoretical consequence. His discoveries proved of immense importance in reducing the incidence of infectious diseases, while his development of the pasteurization process transformed the production and preservation of food and, perhaps more important to French merchants, beer.

In the course of his elegant work, Pasteur also contributed directly to the study of life's origin. By introducing the dictum that no life can occur without prior life, his findings pushed back life's origin to an inconceivably remote time and place. How could anyone make observations or perform experiments to study an event so ancient and inaccessible?

SPECULATION

In 1871, Charles Darwin penned a speculative letter to his friend, the botanist Joseph Hooker. "If (and Oh! What a big if!)," he wrote, "we could conceive in some warm little pond, with all sorts of ammonia and phosphoric salts, light, heat, electricity, etc., present, that a protein compound was chemically formed ready to undergo still more complex changes. . . ." Many contemporary chemists must have agreed that life's origin, wherever and however it occurred, depended on three key resources.

First and foremost, terrestrial life requires liquid water, the essential growth medium for all living things. All living cells, even those that survive in the driest desert ecosystem, are formed largely of water. Our bodies are typically 70 percent water, while many foods are even more water-rich. Surely the first cells arose in a watery environment.

Life also needs a ready source of energy. The radiant energy of the Sun provides the most obvious supply for life today, but bolts of lightning, asteroid impacts, Earth's inner heat, and the chemical energy of certain unstable minerals have also been invoked as life-triggering energy sources.

And, third, life depends on a variety of chemical elements. All known living organisms consume atoms of carbon, oxygen, hydrogen, and nitrogen, with a bit of sulfur and phosphorus and other elements as well. Atoms of these elements combine in graceful geometries to form essential biomolecules.

In spite of the intrinsic importance of the topic, Darwin's private musings about a chemical origin of life (which he never published) flew in the face of conventional theology. "In the beginning, God created" was sufficient for most people, including a majority of the most distinguished scholars of the time. Not until the 1920s did such scientific speculation take a more formal guise.

Most notable among that pioneering modern school of origin theorists was the Russian biochemist Alexander Oparin. In 1922, while still in his twenties and under the scrutiny of the Soviet hierarchy, Oparin elaborated on the idea that life arose from a body of water that gradually became enriched in organic molecules—the so-called "Oparin Ocean" or the "primordial soup." Somehow, he posited, life emerged as these molecules clustered together and self-organized into a chemical system that could duplicate itself. In many other cultures, where religious doctrine colored thinking on origins, these revolutionary ideas would have been seen as heretical, but Oparin's ideas resonated with the materialist philosophical worldview of the Soviet leadership.

Many of Oparin's postulates were echoed in 1929 in the independent ideas of British biochemist and geneticist J. B. S. Haldane. Haldane, himself a Marxist activist, contributed a brief, perceptive (many contemporaries would have said radical) article entitled "The Origin of Life" to the eclectic secularist British periodical *The Rationalist Annual*. He speculated on the production of large carbon-based molecules under the influence of the Sun's ultraviolet radiation. Given such a chemical environment, Haldane envisioned the first living objects as self-replicating, specialized molecules.

Oparin and Haldane offered original and intriguing ideas. More important, their ideas were subject to experimental testing, but for some reason Oparin and his contemporaries didn't try. It wasn't until after World War II, a time of exuberant, can-do scientific optimism, that two chemists demonstrated a systematic experimental approach to prebiotic chemistry.

EXPERIMENTS

The apparently universal requirements of water, energy, and chemicals hint at a simple experimental approach for studying the origin of life. Such landmark experiments were devised early in the 1950s by University of Chicago chemist Harold Urey, a Nobel Prize winner of great renown, and his unknown second-year graduate student, the 23-year-old Stanley Miller.

Jeffrey Bada, a Miller student in the 1960s and an unswerving proponent of his teacher's ideas in subsequent years, reveals some of the behind-the-scenes context of the historic Chicago work in his 2000 book (coauthored with Christopher Wills), *The Spark of Life*. Urey planted the seed for the now classic experiment in the fall of 1951, when he presented a seminar attended by Miller, who was a beginning graduate student. In that lecture and related publications, Urey proposed that life-triggering organic molecules might have been produced in abundance in a plausible primitive atmosphere of hydrogen, methane, and other relatively reactive gases.

A year later, after trying his hand at an unsuccessful project in nuclear theory, Miller asked Urey if he could attempt an experiment with the hopes of synthesizing amino acids, the building blocks of proteins. At first, so the story goes, Urey was reluctant and wanted Miller to try an easier, safer research project. But after continued pressure from Miller, Urey relented and gave his student a limit of one year to make headway.

Scientists revere simple, elegant experiments, and that's exactly what Miller and Urey devised—a primitive Earth sitting on a tabletop in a sealed glass vessel. [Plate 5] To the lower left, Miller positioned a 5-inch-diameter (300-milliliter) flask filled two-thirds with water—the primitive ocean. Glass tubing linked that flask to a 10-inch-diameter (5-liter) gas-filled flask, situated above and to the right. Miller's atmosphere consisted of the relatively reactive gases methane (CH_4, sold commercially as natural gas), ammonia (NH_3, the strong-smelling component of ammonia-based cleaners), and hydrogen (H_2, the explosive lighter-than-air gas of *Hindenburg* fame). Two pointed metal electrodes penetrated the upper flask to simulate "lightning" by sparking, and a flame gently heated the water-filled flask to mimic the ocean's continuous evaporation. Sending sparks through a potentially explosive mix of hydrogen and methane is not for the faint of heart, nor would Miller's unshielded glass apparatus have stood up to today's government safety standards. But it was an elegant experimental design.

Following academic tradition, Miller the student performed the experiments, while his mentor waited in the wings. Miller began by

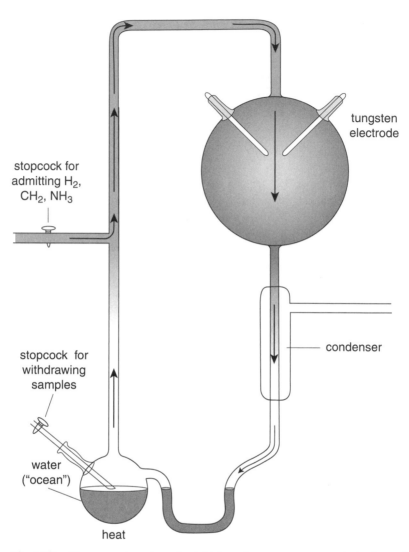

stopcock for admitting H_2, CH_2, NH_3

tungsten electrode

stopcock for withdrawing samples

condenser

water ("ocean")

heat

The Miller–Urey experiment used a tabletop glass apparatus to mimic the early Earth's ocean and atmosphere, while electric sparks simulated lightning. This device rapidly transformed simple gases into essential biomolecules (after Trefil and Hazen, 1992).

pumping out the system to remove any traces of atmospheric oxygen. Then he filled his apparatus with water and a 2:2:1 gas mixture of methane, ammonia, and hydrogen. He heated the water, set off tiny electric sparks in the gases to simulate lightning, and waited.

When the experiment began, the water was pure and clear, but within a couple of days the solution had turned yellowish and a black residue had begun to accumulate near the electrodes. It's easy to imagine Stanley Miller's excitement as he cut his first experiment short to analyze the tantalizing products. Had he produced amino acids?

Organic chemical analysis was no easy chore in the early 1950s. Miller resorted to paper chromatography, a classic and relatively quick technique that separates different molecules into discrete colored spots on absorbent paper. Miller opened the valve and removed the yellow solution. Concerned that microbes might begin to grow in the liquid and confuse his results, he added a lethal dose of mercury chloride to preclude any bacterial contamination. Then, after the tedious routine of drying down and concentrating his sample, he placed a small drop of concentrated run product near one corner of the chromatography paper. It dried as a small yellow dot.

Miller suspended the paper above a narrow trough filled with a freshly made alcohol–water solution. He carefully lowered the paper so that one edge close to the yellow spot just barely dipped into the clear solution, which gradually soaked into the paper and rose up the sheet by capillary action. As the alcohol solution moved, it carried the unknown molecules along with it. Within a few minutes the chemical spot had been smeared out into a narrow, 3-inch-long streak of unknown, mostly invisible chemicals.

Miller let his paper dry, rotated it 90 degrees, and repeated the process with a second solvent, this time a phenol solution. Molecules move differently through paper under the influence of different solvents, so this second step spread the 3-inch streak of chemicals into a diagnostic two-dimensional pattern of spots. After a second drying, Miller's final step was to treat the paper with ninhydrin, a chemical that stains otherwise invisible amino acids into distinctive colors.

Almost immediately a discrete purple spot appeared exactly at the expected position for glycine ($C_2H_5NO_2$), the simplest of life's amino acids. Starting with nothing more than water and a few simple gases, Miller had made one of life's essential biomolecules. The graduate student and his advisor were elated, for the tabletop experiment had worked much faster than they expected.

Miller repeated the entire experiment for a week's duration, cranking up the heat to a slow boil. The results were amazing. The water quickly yellowed, then gradually turned intriguing shades of pink and eventually to a deep red, while black gunk oozed down the sides of the larger flask. This time when Miller analyzed the contents of the flask, he found a complex mixture of organic molecules including at least a half-dozen different amino acids. Reactions of water and air had produced organic molecules in abundance.

Harold Urey encouraged the young investigator to write up his results immediately. Miller obliged and submitted a short manuscript to the high-profile journal *Science* in mid-February 1953, a scant five months since the project's inception. Urey also generously withdrew his name from the paper so that the graduate student, not the Nobelist, would receive the lion's share of credit.

Stanley Miller's first publication was a bombshell. His two-page article in the May 15, 1953, issue of *Science* announced "A Production of Amino Acids Under Possible Primitive Earth Conditions." The press had a field day. The *New York Times* printed a feature story, LIFE AND A GLASS EARTH, while tabloids speculated about synthetic life crawling out of test tubes.

THE COTTAGE INDUSTRY OF PREBIOTIC CHEMISTRY

The Miller–Urey experiment transformed the science of life's chemical origins. For the first time, an experimental protocol mimicked plausible life-forming processes. For decades to come, Miller and his students would dominate the origin-of-life community.

Given such an exciting result, other groups jumped at the chance to duplicate the amino acid feat. Independent confirmation of Miller's claims came within a few years from chemists at the University of Bristol in England and at the Carnegie Institution in Washington, D.C. (A careful search of the scientific literature also turned up a remarkably similar series of experiments that had been conducted decades earlier by the German chemist Walter Löb.) Thousands of subsequent experiments during the past half century have outlined the promise, as well as the limitations, of this idea that life arose as a chemical process at the surface of the Earth. Time and time again, variations of the

Miller–Urey process, including experiments using ultraviolet radiation, different gas mixtures, powdered minerals, and more, have demonstrated the synthesis of life's most basic building blocks. Relatively easy to run, and now a relative cinch to analyze, these experiments continue to be a sort of cottage industry in origin-of-life research circles.

This genre of experiments yields amazing results. Dozens of amino acids have been synthesized from scratch, along with membrane-forming hydrocarbons, energy-rich sugars and other carbohydrates, and metabolic acids. The early inventories even included some of the building blocks of the genetic molecules DNA and RNA, though at first the 5-carbon sugar ribose (the "R" of RNA) and the two purines essential to both, adenine ($C_5H_5N_5$) and guanine ($C_5H_5ON_5$), were notably absent.

Much of this molecular diversity occurs because electric sparks and ultraviolet radiation trigger the formation of highly reactive chemical species, such as hydrogen cyanide (HCN) and formaldehyde (CH_2O), which readily link to other molecules. Miller suspected, for example, that most of the amino acids produced in his experiments arose by a chemical process known as Strecker synthesis, in which hydrogen cyanide reacts with formaldehyde and ammonia.

Enthusiasm grew as other scientists discovered promising new chemical pathways. In 1960, John Oró of the University of Houston turned scientific heads when he discovered that a concentrated hydrogen–cyanide solution, when heated, produced lots of adenine, one of the missing purines and a crucial biomolecule that also plays a role in metabolism. Other chemists conducted similar experiments, starting with relatively concentrated solutions of formaldehyde (CH_2O), a molecule thought to be common in some prebiotic environments. These experiments produced a rich, though random, variety of sugars, including a modest yield of ribose. Gradually, through such specialized experiments, gaps in the prebiotic inventory of life's molecules were filled in.

The experiments of Oró and others, relying as they did on relatively concentrated solutions of reactive organic molecules, raised some eyebrows. The Miller-type spark experiments by themselves had never yielded hydrogen cyanide in sufficient concentrations to produce much adenine, or enough formaldehyde to make much ribose. But discoveries in the mid-1960s in the lab of Salk Institute pioneer Leslie Orgel pointed to a plausible fix. The early Earth was not uniformly hot; it probably had ice caps and may have periodically experienced more extensive "ice ages," as well. Orgel realized that such conditions might promote intriguing organic reactions. When an organic-rich water solution freezes, pure water ice crystals grow, while the dwindling reservoir of residual liquid becomes an increasingly concentrated organic brine (and recall that a higher concentration of "agents" may facilitate emergence).

Orgel and co-workers exploited this idea by slowly freezing flasks of dilute HCN solutions to −20°C. The procedure produced tiny volumes of extremely concentrated hydrogen cyanide, which reacted over weeks to months to produce small linkages of up to four HCN molecules. This curious phenomenon became the inspiration for one of the longest experiments in the history of origins research. Sometime in the mid-1970s, Miller, now a professor at the University of California, San Diego, and his co-workers repeated the Orgel protocol and stored their frozen flasks in the back of a freezer. In the late 1990s, more than two decades later, they removed the frozen solutions, which had developed curious dark concentrated clumps that were rich in organics. Analysis revealed an abundant production of adenine. It takes a lot of time for reactions to proceed at ultracold temperatures, but the primitive Earth had time to spare.

These remarkable results seem to defy convention: Heat, not cold, normally drives chemical reactions. You don't make a cake by freezing batter. Nevertheless, additional freezing experiments have produced amino acids and other interesting biomolecules by this counterintuitive process of concentration. In this curious way, prebiotic cycles of freezing and thawing may have enhanced the emergence of biomolecules and thus provided a pathway to life.

DOUBTS

As exciting and important as the Miller–Urey results may be, seemingly intractable problems remain. Within a decade of Miller's triumph, serious doubts began to arise about the true composition of Earth's earliest atmosphere. Miller and Urey had exploited a highly reactive atmosphere of methane, ammonia, and hydrogen, which seemed a plausible early atmosphere to them. But by the 1960s, new geochemical calculations along with data from ancient rocks pointed to a much less reactive early atmosphere of nitrogen and carbon dioxide, two gases that do almost nothing of interest in a Miller–Urey apparatus.

Miller and his supporters continue to counter with a pointed argument, difficult to dismiss. Life's biomolecules match those of the original Miller–Urey experiment with great fidelity, they say. Doesn't that fact alone argue for an atmosphere rich in reactive methane? Harold Urey is said to have often quipped, "If God did not do it this way, then He missed a good bet." Nevertheless, most geochemists now discount the possibility of more than a trace of atmospheric methane or ammonia at the time of life's emergence.

Added to this atmospheric concern is the fact that the molecular building blocks of life created by Miller and his colleagues represent only tiny steps on the long road to life. Living cells require that such small molecules be carefully selected and then linked together into vastly more complex structures—cell membranes, protein catalysts, DNA, RNA, and other so-called macromolecules. The prebiotic ocean was an extremely dilute solution of many thousands of different organic molecules, most of which play no known role in life. By what emergent processes were just the right molecules selected and organized?

The Miller–Urey scenario suffers from yet another nagging problem. Macromolecules tend to fragment, rather than form, when subjected to the energetic insults of lightning and the Sun's ultraviolet light. These so-called ionizing forms of energy are great for making reactive molecular fragments that combine into modest-sized molecules like amino acids. Combining many amino acids into an orderly chainlike protein, however, is best accomplished in a less destructive energy domain. Emergent complexity relies on a flow of energy, to be sure, but not too much energy. Could life have emerged in the harsh glare of daylight, or was there perhaps a different, more benign origin environment?

Faced with such an impasse, a few maverick scientists began to look at other plausible venues for the cradle of life.

7

Heaven or Hell?

It is we who live in the extreme environments.
Thomas Gold, *The Deep Hot Biosphere*, 1999

For centuries the primary source of life's energy has been as well established as any precept in biology. Every high-school textbook proclaims what we all have accepted as intuitively obvious: All life depends ultimately on the Sun's radiant energy. Nor has there been reason to doubt that claim until recently. But new discoveries of deep life—life-forms at the darkest ocean depths and microbes buried miles beneath Earth's surface in solid rock, forever beyond the Sun's influence—have toppled this comfortable certainty.

If science has taught us anything, it's that cherished notions about our place in the natural world often turn out to be dead wrong. We observe that the Sun rises in the morning and sets at night. An obvious conclusion, reached by almost all observers until relatively recently in human history, is that the Sun circles the Earth. Yet we now know that sunrise and sunset are consequences of Earth's rotation; Earth orbits the Sun, and we are not at the physical center of the universe. We observe mountains and oceans as grand unchanging attributes of the globe—on the scale of a human life, these features are for all intents and purposes permanent. Yet we have learned that through the inexorable processes of plate tectonics, every topographic feature on Earth is transient over geological time and that our war-contested political boundaries are destined eventually to disappear.

The great power of science as a way of knowing is that it leads us to conclusions about the physical universe that are not self-evident. Repeatedly, the history of science has been punctuated by the overthrow of the obvious. Could our intuitive view of life's original energy source be in error as well?

ENERGY

All living cells require a continuous source of energy. Without energy, organisms cannot seek out and consume food, manufacture their cellular structures, or send nerve impulses from one place to another. Lacking energy, they cannot grow, move, or reproduce. Reliable energy input is also essential to maintain the genetic infrastructure of cells, which are constantly subjected to damage by nuclear radiation, toxic chemicals, and other environmental hazards.

Metabolism, the means by which organisms obtain and use energy, is an ancient chemical process that takes place in every living cell, including all of the tens of trillions of cells in our bodies. Until recently, scientists claimed that the metabolic pathways of virtually all life-forms rely directly or indirectly on photosynthesis. At the base of the food web, we find plants and a host of one-celled organisms that use the Sun's light energy to convert water and carbon dioxide into the chemical energy of sugar molecules (carbohydrates) plus oxygen. Plants manufacture carbohydrates, such as the starch of potatoes and the cellulose of celery, to build leaves, stems, roots, and other physical structures. They also process sugar molecules to provide a source of chemical energy that powers the plant cell's molecular machinery.

While plants synthesize their own carbohydrates, animals and other nonphotosynthetic life higher up the food chain must find another source of sugar. That's why we eat plants, or eat animals that eat plants. Plants synthesize sugar molecules and oxygen from water plus carbon dioxide. Our bodies convert sugar molecules along with the oxygen we breathe to produce water plus the waste gas carbon dioxide. There's an elegant chemical symmetry to this story; the biological world seemed much simpler when the Sun was life's only important energy source.

DEEP ECOSYSTEMS

Our view of life on Earth changed forever in February 1977, when Oregon State University marine geologist Jack Corliss and two crewmates guided the submersible *Alvin* to the deep volcanic terrain of the East

Pacific Rise, 8,000 feet down. This undersea ridge off the Galápagos Islands was known to be a zone of constant volcanic activity associated with the formation of new ocean crust. Oceanographers have documented thousands of miles of similar volcanic ridges, including the sinuous Mid-Atlantic Ridge that bisects the Atlantic Ocean—the longest mountain range on Earth.

On this particular dive, just one of hundreds that *Alvin* had logged, the scientists hoped to locate and examine a submarine hydrothermal vent, a kind of submarine geyser where hot water jets upward into the cool surrounding ocean. What Corliss and crew discovered was a vibrant and totally unexpected ecosystem with new species of spindly albino crabs, football-sized clams, and bizarre 6-foot tubeworms. One-celled organisms also abounded, coating rock surfaces and clouding the water. These communities, thriving more than a mile and a half beneath the sea, never see the light of the Sun.

In these deep undersea zones, microbes serve as the primary energy producers, playing the same ecological role as plants do on Earth's sunlit surface. These one-celled vent organisms exploit the fact that the cold oxygen-infused ocean water, the hot volcanic water, and the sulfur-rich mineral surfaces over which these mixing fluids flow are not in chemical equilibrium. This situation is similar to the disequilibrium between a piece of coal and the oxygen-rich air. Just as you can heat your house or power machinery by burning coal (thus combining unstable carbon and oxygen to make stable carbon dioxide), so too can these deep microbes obtain energy by the slow alteration of unstable minerals.

The unexpected discovery of this exotic ecosystem was news enough, but Corliss and his Oregon State colleagues soon tried to push the story further. They saw in the vents an ideal environment for the origin of life. Details of this story have become clouded by more than 20 years of sometimes revisionist history. Corliss claims the idea for himself: "I began to wonder what all this might mean, and this sort of naïve idea came to me," he told an interviewer more than a decade later. "Could the hydrothermal vents be the site of the origin of life?" [Plate 5]

A different history emerges from others close to the story. According to John Baross, a former faculty colleague of Corliss and an expert on microbes in extreme environments, the hydrothermal vent theory of life's origin was first proposed and developed by a perceptive Oregon State graduate student named Sarah Hoffman. She wrote the basic outlines of the hypothesis in 1979, as a project for a biological oceanography seminar taught by Charles Miller, another OSU oceanographer. Hoffman, in frequent consultation with Baross, developed the novel idea as it would appear in print. The two of them claim that the more senior Corliss seized the paper as his own, allowing them, as his coauthors, to expand and polish the prose to conform to the conventions of scientific publishing, after which he submitted the work and placed his name first on the author list. With three coauthors—Corliss, Baross, and Hoffman—the paper would forever be known as "Corliss et al." Corliss would get the fame, while Hoffman and Baross were effectively relegated to footnote status.

Whoever deserves the credit, the hydrothermal-origins thesis is elegantly simple and correspondingly influential. Modern organisms do, in fact, thrive in deep hydrothermal ecosystems. Fossil microbes recovered from 3.5-billion-year-old hydrothermal deposits reinforce this observation. Even without the energy of sunlight, nutrients and chemical energy abound in hydrothermal systems. The OSU scientists saw hydrothermal systems as "ideal reactors for abiotic synthesis," and they proposed a sequence of chemical steps for the potentially rapid emergence of life.

The controversial manuscript was not eagerly received; it bounced around for the better part of a year. First it was rejected by *Nature*, then by *Science*. At the time, Stanley Miller and his protégés dominated the origin-of-life research game, which had seen more than its fair share of quacks and crackpot theories. They were not about to let such unsupported speculation sully their field. Hydrothermal temperatures were much too hot for amino acids and other essential molecules to survive, they said. "The vent hypothesis is a real loser," Miller complained to a reporter for *Discover* magazine. "I don't understand why we even have to discuss it."

Miller's followers found other good reasons to attack the paper. Corliss and co-workers had the ancient ocean chemistry all wrong, they said. Modern hydrothermal ecosystems rely on oxygen-rich ocean water, whose composition is an indirect consequence of plants and photosynthesis. The prebiotic ocean would not have been oxygen-rich, so the proposed life-sustaining chemical reactions would have proceeded slowly, if at all. The bottom line? Decades of Miller-type experiments confirm what is intuitively obvious: Life began at the surface, so why confuse the issue?

Eventually the Corliss, Baross, and Hoffman manuscript was published, in a supplement to the relatively obscure periodical *Oceanologica Acta*, a journal that not one in a hundred origin-of-life researchers would see. Nevertheless, good ideas have a life of their own, and copies of the paper, entitled "An Hypothesis Concerning the Relationship Between Submarine Hot Springs and the Origin of Life on Earth," began circulating. I have seen dog-eared underlined photocopies of copies of copies on several colleagues' desks, and I have a pretty battered copy of my own.

New support for the idea gradually consolidated, as hydrothermal ecosystems were found to be abundant along ocean ridges in both the Atlantic and Pacific. It was realized that at a time when Earth's surface was blasted by a continuous meteorite bombardment, deep-ocean ecosystems would have provided a much more benign location than the surface for life's origin and evolution. New discoveries of abundant primitive microbial life in the deep continental crust further underscored the viability of deep, hot environments. By the early 1990s, the deep-origin hypothesis had become widely accepted as a viable, if unsubstantiated, alternative to the Miller surface scenario.

Of the three authors, only John Baross remains active and influential in the field. In 1985, he accepted a professorship at the University of Washington, where he has developed a leading research program on hydrothermal life. His work on deep-sea-vent microbes, often in collaboration with his wife, Jody Deming, who is also a professor of oceanography at the University of Washington, has placed Baross at the forefront of the highly publicized research field of "extremophile" microbes. Sarah Hoffman's graduate work in geochemistry was interrupted by illness, and, after her recovery, she pursued a singing career. As for Corliss, always a bit idiosyncratic, in 1983 he left Oregon for the Central European University in Budapest, where he worked briefly on the deep-origins hypothesis, but soon took up research in the more abstract field of complex systems. After a 3-year stint as director of research at the controversial Biosphere 2 environmental station in Arizona, he returned to Budapest, having abandoned studies of the deep ocean.

LIFE IN ROCKS

Following the revolutionary hydrothermal-origins proposal, numerous scientists began the search for life in deep, warm, wet environments. Everywhere they looked, it seems—in deeply buried sediments, in oil wells, even in porous volcanic rocks more than a mile down—microbes abound. Microbes survive under miles of Antarctic ice and deep in dry desert sand. These organisms appear to thrive on mineral surfaces, where interactions between water and chemically unstable rocks provide the chemical energy for life.

One of the most dramatic and difficult pursuits involves deep drilling for life in solid rock. The oil industry has perfected the practice of deep drilling, thanks to decades of experience and vast infusions of cash. They can penetrate several miles into the Earth, drill at angles and around obstacles, and cut through the hardest known rock formations in their quest for black gold. So the problem for geoscientists looking for microbes a mile or more down isn't how to get there, it's how to get there without contaminating the drill hole with hoards of surface bugs. Bacteria are everywhere—in the air, in the water, and in the muck used to lubricate and cool diamond drill bits as they cut through layers of rock. It's relatively easy to bring up rock cores from a couple of miles down, but those slender cylinders of rock will have already been exposed to surface life by the drilling process. What to do?

The commonest retrieval trick is to add a colorful dye or other distinctive chemical tracer to the lubricating fluid. When drillers extract a deep core, it becomes obvious whether or not the rock has been contaminated in the process. Porous sediments or highly fractured formations soak up the dye and thus prove unsuitable for analysis, but many rocks turn out to be impermeable and thus are ideal for recovering deep life.

The search for subsurface microbes began in earnest in 1987, when the Department of Energy decided to drill several 500-meter-deep boreholes in South Carolina near the Savannah River nuclear processing facility. As cores were brought to the surface, drillers quickly isolated them in a sterile plastic enclosure with an inert atmosphere. Researchers then cut away the outer rind of the drill core to reveal pristine rock samples, which were shipped to analytical facilities across the country. The results were spectacular. The deep South Carolina sediments were loaded with microbes that had never seen the light of day.

Subsequent drilling studies have revealed that microbes live in every imaginable warm, wet, deep environment—in granite, in basalt on land and basalt under the ocean, in all variety of sediments, and also in metamorphic rocks that have been altered by high temperature and pressure. Anywhere you live, drill a hole down a mile and the chances are you'll find an abundance of microscopic life.

MINING FOR MICROBES

Earth's deep mining and tunneling operations provide the new breed of geobiologist with an invaluable complement to drilling. Mine tunnels have the advantage that researchers can visit microbial populations in their native habitat. Earth's deepest mines, the fabled gold mines of South Africa's Witwatersrand District, have thus become the site of the heroic and potentially dangerous efforts of Princeton geologist Tullis Onstott.

The East Drietontein Mine, located about 60 miles southwest of Johannesburg, is a vast network of underground workings, reaching more than 2 miles into the crust. A small army of miners labor around the clock for gold. Despite one of the largest air-conditioning systems in the world, these deep tunnels remain at an oppressive 140°F from the heat of Earth's interior, while air pressure is twice that of the surface. Onstott learned the dangers of the place on his first descent: "It was 'Don't step there, don't touch that,'" he told a writer for the *Princeton Weekly Bulletin*. "All I knew is that it was deep and dark and hot."

Every so often, as miners blast new adits, a small flow of water appears—groundwater that has spent countless thousands of years filtering down from the surface and has accumulated in small cracks and fissures, nurturing a tenuous ecosystem of microbial life. Onstott's team, typically a half-dozen young and hearty students and postdocs, camp at the surface with a functional array of sterile sample-collection hardware at hand. When news of a fresh water flow comes in, they scramble to the site, though the miles of elevators and tunnels can take almost an hour to traverse. They have to work fast, both to avoid disruption of the mining routine and because prolonged exposure to the hellish conditions can kill them.

They photograph the site, record its location and geological setting, and collect as many gallons of water as possible fresh from the point of flow. They benefit from the seep's positive water pressure, which prevents much back contamination from the miners' activities or their own collection efforts. Exhausted and sweating profusely, they lug the heavy water-filled bottles to the surface for further investigation.

Remarkably, every single sample from Earth's deepest mines holds microbes that have never seen light, surviving on a meager supply of underground chemical energy. Such deep life lives at a sluggish pace that defies our experience. Isotopic measurements reveal that a single cell may persist for thousands of years, "doing" almost nothing before dividing into two. Colonies of organisms commonly remain isolated from the surface for millions of years. So tenuous are the chemical resources of these deep rocks that reproduction and growth are luxuries seldom indulged. By the same token, deep rocks provide an unvarying safe and reliable environment: no predators, no surprises—unless of course a miner happens to blast into your rocky home of a million years!

THE DEEP HOT BIOSPHERE

The abundance of subterranean one-celled creatures, thriving far from the light of the Sun, inspires the imagination and hints at novel scenarios for life's origin. Of all the scientists in pursuit of deep life, none displayed greater imagination than the late brilliant and pugnacious iconoclast Thomas Gold.

Austrian-born Tommy Gold began his scientific career as an astrophysicist in Britain, but in 1959 he was lured to Cornell University to head the Center for Radiophysics and Space Research. He would achieve lasting scientific fame with his inspired theory that pulsars, steady pulsating radio sources discovered in 1967, are actually rapidly rotating neutron stars. Many honors, including election to the Royal Society of London and the National Academy of Sciences, soon followed.

Most scientists would have been content to excel in one chosen area, but Gold throughout his career repeatedly ventured into new and controversial academic domains. In the 1940s, he conducted experiments on hearing and the structure of the mammalian inner ear. Speculative papers on dramatic instabilities of Earth's rotation axis, on steady-state cosmological models of the universe, and on the potential danger to astronauts of deep powdery lunar soils peppered his lengthy curriculum vitae.

In 1977, Gold, by then a safely tenured professor at Cornell, rattled the well-established field of petroleum geology. Geologists had long declared that petroleum is a fossil fuel, formed when huge quantities of decaying cells accumulate over millions of years, to be buried and processed by Earth's heat and pressure. The evidence is overwhelming: Petroleum occurs in sedimentary layers that once held abundant life; petroleum is rich in distinctive biological molecules; petroleum's carbon isotopes also point to a biological source. Armed with these and a dozen other lines of evidence, the case for fossil fuels was open and shut.

Gold disagreed. Petroleum holds lots of distinctive biomolecules, to be sure, but oil fields also contain, along with methane, an abundance of helium gas—a light gas that quickly escapes into space and so could only come from deep within the Earth. How could one reconcile the mixing of deep helium with surface biology? Gold's conclusion: The organic molecules that eventually become petroleum are produced deep underground by purely chemical processes and are then modified by the action of subsurface microbes.

In this scenario, vast sources of primordial hydrocarbons—the major molecular components of oil—exist in Earth's mantle. Because they are lighter than the surrounding rocks, these hydrocarbons slowly but surely rise toward the surface, constantly refilling petroleum reservoirs. In Gold's heretical view, oil is thus a renewable resource instead of a finite one built up over millions of years by the burial and decay of once-living cells. In an extraordinary move, Gold first published this novel idea as an op-ed piece in the June 8, 1977, issue of the *Wall Street Journal*. An oil-hungry nation in the midst of an energy crisis took considerable notice of the radical hypothesis.

A scientific theory is useful only if it is testable, and Gold soon proposed a test of dramatic proportions. Gold's oil-from-below hypothesis predicts that great oil fields should arise equally in many different types of rock, but all known petroleum has been found in layers of exactly the kind of sedimentary formations that would have collected abundant remnants of past life. Gold countered, logically, that petroleum geologists never look for oil anywhere but in those sedimentary formations. Perhaps, he suggested, immense new oil fields were waiting to be found in igneous and metamorphic rock.

Armed with his provocative theory and a persuasive, dynamic oratorical style, he presented a simple (and expensive) proposal to the Swedish State Power Board in 1983. Drill an oil well in solid granite—the last place on Earth a petroleum geologist would look. He had targeted a unique and tempting granitic mass, the Siljan Ring impact site in central Sweden. This highly fractured granite body, formed 368 million years ago when an asteroid shattered the crust, holds tantalizing hints of petroleum in the form of carbon-filled cracks and flammable methane seeps. Gold's enticing rhetoric, amplified by increasing optimism from Dala Deep Gas, the company formed to do the drilling, lured energy-poor Swedes into spending millions of dollars on exploratory holes.

Seven years and $40 million later, a 6.8-kilometer-deep hole had produced only modest amounts of oil-like hydrocarbons and methane gas—a small enough yield for oil experts to say "I told you so," but large enough to convince Gold that his theory was right. Where else, he asked, could that trace of organic molecules have come from? Nevertheless, most scientists saw the Siljan experiment as a failure, and no one is likely to drill for oil in granite again, at least anytime soon.

The modest production of methane gas and smelly, oily sludge from the Swedish wells inspired Gold to elaborate on his theory. Extrapolating far beyond his peers, he described "the deep hot biosphere" in several articles and a popular 1998 book of that title. The vast, deep hydrocarbon reserves at the heart of Gold's 1977 hypothesis provide a wonderful food source for deep microbes, which coincidently leave their biological overprint on the otherwise abiotic oil. In entertaining prose, he reviewed the growing and accepted body of evidence for deep life in many types of rock—all very reasonable stuff. Gold's conclusion: Deep microbial life, much of it nourished by upwelling hydrocarbons, accounts for fully half of Earth's total biomass. Though living cells represent a tiny fraction of Earth's total rock mass, the volume of rock is so vast—a few billion cubic kilometers—that it shelters astronomical numbers of microbes. Inevitably, our view of life has been skewed because these microscopic life-forms lie completely hidden from everyday view.

In April 1998, I invited Gold to visit the Geophysical Lab and present his ideas at our regular Monday morning seminar. Seemingly unfazed by more than two decades of impassioned objections to his views, he delivered a polished and forceful account of many lines of evidence that petroleum is abiotic and rises from the depths. Geology, biology, thermodynamics, experiments on organic molecules, carbon isotopes, observations of diamonds, and, of course, the chemical properties of petroleum itself all came into play. Knowing our special interest in life's origins, he underscored the possible role of this deep hydrocarbon source in supplying critical molecules for prebiotic processes. Perhaps, he posited, life arose from those deep sources of organic molecules. Throughout the entertaining lecture, he bolstered his controversial conclusions with rhetorical flourishes more suited to a courtroom than a scientific seminar ("the only possible explanation," "no question remains," the evidence "persuades one completely," and the like).

I doubt that anyone was persuaded completely, but we had to admire his creativity and conviction. And Tommy Gold helped to remind us all, once again, how much we don't know about the interior of our planet just a few miles beneath our feet.

HEAVEN VERSUS HELL

If so many organisms exist beyond the Sun's radiant reach, then geothermal energy, and the abundant chemically active mineral surfaces that are synthesized in geothermal domains, must be considered as a possible triggering power source for life. To be sure, sunlight remains the leading contender for life's original energy source. The vast majority of known life-forms do rely, directly or indirectly, on photosynthesis. In many scientific circles, a surface origin of life in a nutrient-rich ocean, under a bright Sun, remains the seemingly unassailable conventional wisdom.

But that nagging problem of macromolecular formation remains. Most known living species depend on the Sun directly or indirectly, but the Sun's harsh ultraviolet radiation inhibits the emergence of the larger multimolecular structures on which all organisms depend. Furthermore, if the earliest life of almost 4 billion years ago was confined to the sunlit surface, how did it escape the brutal, sterilizing final stages of bombardment by asteroids and comets? As Gold said over and over again, Earth's surface, bathed in solar radiation and blasted by lightning, is the truly extreme environment.

And there's another reason to look closely at the possibility of hydrothermal origins. If life is constrained to form in a sun-drenched pond or ocean surface, then Earth, and perhaps ancient Mars or Venus, are the only possible places where life could have begun in our solar system. If, however, living cells can emerge from deeply buried wet zones, then life may be much more widespread than previously imagined. The possibility of deep origins raises the stakes in our exploration of other planets and moons.

Jupiter's fourth largest moon, Europa, presents a particularly promising target for exploration. According to recent observations, Europa is covered with a veneer of ice about 10 kilometers thick, covering a deep ocean of liquid water. Hydrothermal activity on the floor of that ocean might be an ideal environment for life-forming chemistry. Saturn's largest moon, Titan, is another intriguing world. Though much colder than Earth, Titan has an organic-rich atmosphere slightly denser than Earth's and, like all large bodies, its own sources of internal heat energy.

The idea that life may have arisen in a deep, dark zone of volcanic heat and sulfurous minerals flies in the face of deeply ingrained religious metaphors. To many people, the Sun represents the life-giving warmth of heaven, while sulfurous volcanoes are the closest terrestrial analog to hell. How could life have come from such a dark, hostile environment?

Nature is not governed by our metaphors, however cherished they may be. Life as we know it demands carbon-based chemicals, a water-rich environment, and energy with which to assemble those ingredients into a self-replicating entity. Ongoing laboratory experiments that simulate deep conditions as well as those on the surface—coupled with observations of environments elsewhere in the solar system—will be the ultimate arbiters of truth.

8

Under Pressure

Where is this original homestead of life? . . . a place where hot
volcanic exhalations clash with a circulating hydrothermal
water flow; a place deep down where a pyrite-forming
autocatalyst once gave, and is still giving, birth to life.
Günter Wächtershäuser, 1988

The first emergent step in life's chemical origin, wherever it may
have occurred, must have been the synthesis of organic molecules.
Stanley Miller's experiments amply demonstrated one potential ancient energy source with which to effect this synthesis—lightning near
the ocean–atmosphere interface. Subsequent workers, notably Thomas
Gold, postulated deeper origins and chemically derived energy sources.
Any plausible source of carbon-based molecules should be fair game
for origin researchers. Every plausible source deserves rigorous study.

In 1996, when George Cody, Hat Yoder, and I began our research
on hydrothermal chemistry and the origins of life, we were novices in
the origins game, too naïve to suspect that we were wasting our time.
We thought that our experimental studies of organic synthesis in hot,
high-pressure water provided compelling evidence that hydrothermal
zones could have played a role in the prebiotic production of life's molecules. Stanley Miller and his students, most notably chemist Jeffrey
Bada at the Scripps Institution of Oceanography, would have been
happy to set us straight, but we had yet to cross paths with them.
Harold Morowitz, whose ideas instigated the research, had warned us
that we might meet with some criticism. He was right.

It must have seemed odd to the scientific community that a group
of geologists and geophysicists was studying the origin of life. For al-

most a century, our home base, the Carnegie Institution's Geophysical Laboratory, had supported research on the physics and chemistry of Earth materials at extreme temperatures and pressures. Carnegie scientists had melted granite, synthesized basalt, and heated and squeezed just about every known rock-forming mineral to understand how our dynamic planet works. They routinely achieved temperatures of thousands of degrees and pressures of millions of atmospheres—extreme environments that would seem to have little to do with life's delicate chemistry. Nevertheless, the lab's arsenal of furnaces and pressure vessels stood ready to heat and squeeze almost any sort of sample, and experiments on carbon compounds weren't intrinsically different from any other chemical we might have chosen to study.

EXPERIMENTS

Now, more than 2,000 experiments later, our hydrothermal organic-synthesis program is expanded, well funded by NASA, and going strong. Our simple strategy, much in the spirit of Stanley Miller's original experiments, has been to subject plausible prebiotic chemical mixes of water, carbon dioxide, and minerals to a controlled environment—in our case, a high-pressure and high-temperature environment typical of deep-ocean hydrothermal systems—still often employing the classic gold-tube protocol described in the Prologue.

Our first experiments (the ones with pyruvate and carbon dioxide) were exhilarating. In that series we synthesized sugars, alcohols, and a host of larger molecules, many incorporating dozens of carbon atoms. Under extreme hydrothermal conditions, the pyruvate molecules polymerize; in some cases they form rings and branching structures reminiscent of molecules found in modern cells.

In a later set of experiments, we sealed water, nitrogen gas, and powdered minerals into the gold tubes—a reasonable proxy for primitive conditions at some hydrothermal vents. In these experiments, we produced ammonia, an essential starting material for synthesizing amino acids and other biomolecules. Our work suggests that hydrothermal vents may have been one of the principal sources of the early Earth's ammonia. Carrying on with this line of experiments, we put ammonia into a capsule with pyruvate and found that these two compounds react to form the amino acid alanine. By 1999, we had estab-

Plate 1

Harold Morowitz, standing next to his simplified metabolic chart. (Courtesy of H. Morowitz)

George Cody calibrates his gas chromatography mass spectrometer, or GCMS, prior to an experiment. (R. Hazen)

Hatten S. Yoder Jr. with his high-pressure "bomb." (Carnegie Institution of Washington)

Plate 2

Circles of stones form spontaneously in Arctic regions due to the action of freeze-thaw cycles. (Kessler and Werner 2003; © AAAS)

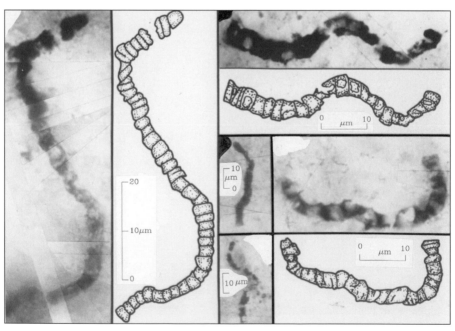

The disputed 3.45-billion-year-old Apex Chert fossils of J. William Schopf. (Schopf 1993; © AAAS)

Plate 3

Bill Schopf (right) peering at Martin Brasier (at the lectern) during their April 2002 debate at the NASA Astrobiology Institute meeting at Moffett Field, California. (O. Green)

Chris Hadidiacos at the Geophysical Laboratory's electron microprobe. (R. Hazen)

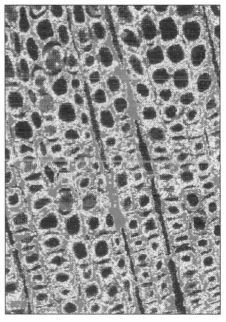

Electron microprobe image of cellular structures from a 400-million-year-old fossil wood (width = 0.3 mm). (R. Hazen)

Plate 4

Marilyn Fogel in the isotope lab. (R. Hazen)

Andrew Steele at the laser Raman spectrometer. Jake Maule in the astrobiology lab. (R. Hazen)
(R. Hazen)

Plate 5

Stanley Miller and his electric spark apparatus. (J.E. Strick and S.L. Miller)

A black smoker with associated fauna. (NOAA)

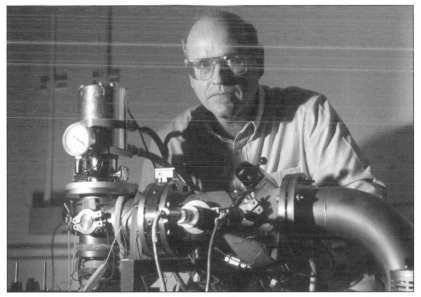

Louis Allamandola at the NASA Ames Research Center with his cryo-vacuum chamber. (Volker Steger/Science Photo Library)

Plate 6

Dave Deamer and Nick Platts in the Santa Cruz lab. (D. Deamer)

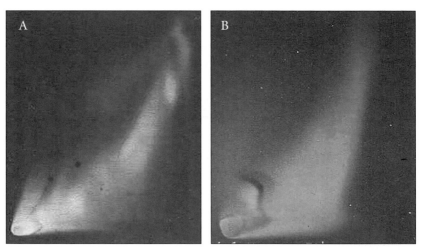

TLC plate of (A) Murchison organics and (B) pyruvate products. Fluorescent patterns on these plates are strikingly similar. (D. Deamer)

Plate 7

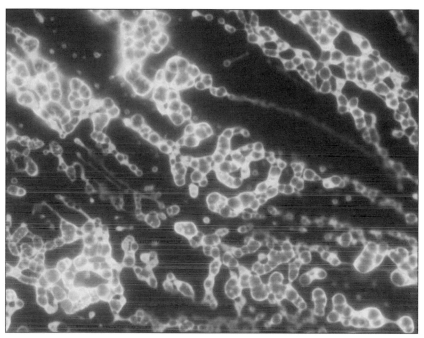

These fluorescent vesicles form spontaneously from products of pyruvate at high temperature and pressure (width = 0.55 mm). (D. Deamer)

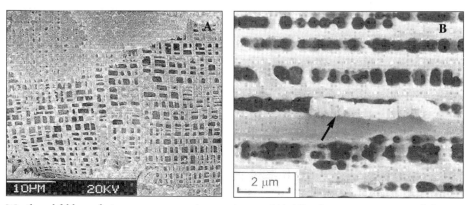

Weathered feldspar features numerous microscopic pores (A). Microbes can occupy some of these cavities (B). (J. Smith)

Plate 8

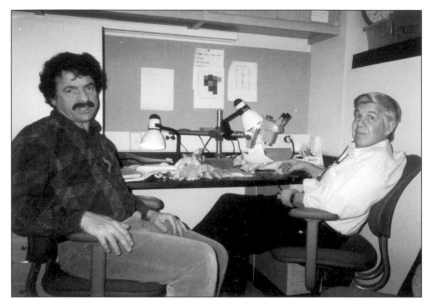

Glenn Goodfriend (left) and Stephen Jay Gould in Glenn's lab at George Washington University, March 2002. (R. Hazen)

Günter Wächtershäuser, who proposed the elaborate Iron–Sulfur World hypothesis for life's origins. (A. Neves, Munich)

Jack Szostak in his Harvard University lab. (J. Szostak)

lished one plausible chemical pathway leading from simple carbon compounds, nitrogen gas, and common minerals to key biomolecules.

In 1997, shortly after our initial exciting results, Cody, Yoder, and I submitted our first papers for peer review. We believed optimistically that our demonstrations of facile hydrothermal organic synthesis would be welcomed as a logical extension of the Miller–Urey approach. After all, we, like Miller a half-century before, had subjected a closed vessel containing water and simple carbon-containing molecules to heat. We had followed Miller's lead in attempting to simulate a plausible prebiotic environment. And we, too, had found an abundance of synthetic organic molecules. We thought the Miller camp might actually welcome our contribution. Boy, were we wrong!

Miller and his scientific cohort had staked their claim to a surface origin of life, and they seemed determined to systematically head off dissenting opinions. Their reasons were straightforward and clearly argued. While organic synthesis may occur near the 100°C limit of liquid water, they argued, at the much higher temperatures of our experiments, the biomolecules of interest, notably amino acids, decompose at a higher rate than they can be manufactured. Hydrothermal zones on the early Earth would have destroyed many essential biomolecules in the prebiotic ocean much faster than they could have been replenished. Such hostile zones were antithetical to life. We weren't the first researchers to suffer these criticisms: The Oregon State group who first outlined the hypothesis of a hydrothermal origin of life following deep-sea discoveries in *Alvin* had found themselves repeatedly thwarted by this argument, their manuscript rejected time and again. However, we were not about to follow their example by publishing in obscure conference proceedings.

Stanley Miller had been relentless in his offensive ever since the hydrothermal-origins idea arose. In a 1988 article in *Nature*, Miller and Jeffrey Bada wrote, "This proposal is based on a number of misunderstandings concerning the organic chemistry involved. . . . The high temperatures in the vents would not allow synthesis of organic compounds, but would decompose them." Miller was more blunt in a *Discover* feature in 1992, when he called the vent hypothesis "a real loser."

In *The Spark of Life*, their popular book on origin theories, Bada and Christopher Wills dubbed those of us working on the hydrothermal hypothesis "ventists," making us sound a little like the members of some fanatical religious cult. Their presentation is engaging—and generally accurate, but like an argumentative rondo the old refrain appears again and again: "Vents are more likely to destroy organic compounds than create them." "This high temperature rapidly destroys organic compounds rather than synthesizing them." "At these high temperatures, amino acids undergo total decomposition on a time scale of only a few minutes." "Vents . . . do terrible things to the primordial soup." Piled onto this catchy litany were other criticisms: Our starting solutions were much too concentrated, the resulting products were much too dilute, and our geological assumptions about vent-water circulation were incorrect.

Our immediate reaction to the assault was defensive and angry. We saw the Miller camp as self-serving, defending old ideas while discounting new possibilities. Such turf-protecting tactics are not unknown in science, but we were more than a little dismayed by such focused efforts to thwart what seemed an interesting and testable idea. Our sense of a conspiracy arrayed against us was heightened when anonymous reviewers recommended rejecting one of our grant proposals on the basis of familiar, party-line criticisms.

The most maddening aspect of Stanley Miller's criticisms wasn't his persistence or even the public way he voiced them. The most maddening aspect was that he was basically right. Our first experiments were admittedly flawed. Our starting mixtures of pyruvate were much more concentrated than any plausible prebiotic solution; indeed, unstable pyruvate is unlikely as an important prebiotic reactant in any environment. What's more, we couldn't control the acidity inside our gold capsule (acidity can drastically alter reaction pathways), we couldn't be sure that all our products formed at high temperature (as opposed to during cooling), and we hadn't proved that the gold was an inert container (might the gold have triggered the observed reactions?). In short, it was a typical start to a complex scientific research program.

The art of science isn't necessarily to avoid mistakes; rather, progress is often made by making mistakes as fast as possible, while avoiding making the same mistake twice. After several hundred experiments under a wide range of pressure and temperature conditions, we began to realize that water, carbon dioxide, and nitrogen are not sufficient by themselves to synthesize very many useful biomolecules. Under the right circumstances we could trigger some interesting organic reactions, but many key biomolecules were missing from the inventory of our products; and the ones we did manage to make were indiscriminate, with numerous molecules of no possible use to biology. We needed some other chemical ingredient, or ingredients, and so our research led us back to my specialty—mineralogy. Minerals, it turns out, have the potential to promote key chemical reactions, while protecting the products from the ravages of high temperature.

Before biology, the only raw materials our planet had to play with were the ocean, the atmosphere, and rocks. It now appears that rocks and minerals, in roles as varied as containers, templates, catalysts, and chemical reactants, may provide a key to understanding life's ancient origins.

ENTER WÄCHTERSHÄUSER

The view of minerals' roles in the origin of life came to the fore with the publication of a seminal paper in 1988: "Before enzymes and templates: Theory of surface metabolism," by the brilliant, combative Günter Wächtershäuser. Wächtershäuser, a chemist by training but a Munich patent attorney by day, erected a sweeping theory of organic evolution in which minerals—mostly iron and nickel sulfides, which abound at deep-sea volcanic vents—provided catalytic, energy-rich surfaces for the synthesis and assembly of life.

Among the many striking aspects of Wächtershäuser's grand scheme is the unusual extent to which he roots his theory in the philosophy of science. Most researchers tend to ignore science philosophers, whose work often seems removed from the practical day-to-day details of experimentation. But Wächtershäuser is an avowed disciple (and a personal friend) of the influential philosopher of science Karl Popper, who encouraged the new theory's development. "During breakfast I mentioned my ideas on an alternative and he urged me to work it out," Wächtershäuser told a *New York Times* reporter not long ago.

According to Popper's dictum, theories, to be theories at all in the scientific sense, must make testable predictions that might be proved false by empirical tests. A theory that makes no testable predictions is pseudoscience according to Popper. By the same token, the more precise (and, therefore, the more falsifiable) its predictions, the more believable a theory becomes when it survives scrutiny.

Most origins researchers before Wächtershäuser had adopted an ad hoc, trial-and-error approach: They cooked up likely geochemical recipes to see what worked, then patched together a theory around those observations. Stanley Miller's experiment made amino acids, so his followers became convinced that that's how it happened in nature. "Ventists," who found primitive life in hydrothermal zones, often fell into the same trap. Theories flourished around particular sets of observations rather than grand falsifiable frameworks. By contrast, Wächtershäuser set out to formulate his theory from scratch, starting with a few plausible assumptions about the nature of the first living entity.

Assumption 1. Reject the "primordial soup" concept. In Wächtershäuser's model, random prebiotic synthesis plays no essential role in life's origin. He argues that the prebiotic emergence of biomolecules à la Miller–Urey was irrelevant because the primordial soup was too dilute and contained too many molecular species that could not contribute to life. It takes a lot of nerve to throw out decades of research, but that's Wächtershäuser's style.

Assumption 2. The first life-form made its own molecules. Most other workers assume that the first living entity scavenged amino acids, hydrocarbons, and other useful molecules from its surroundings—a strategy called heterotrophy (from the Greek for "other nourishment"). Wächtershäuser denies that possibility, since the soup was so dilute and unreliable. He counters that the first life must have been autotrophic ("self-nourishing"), manufacturing its own molecular building blocks from scratch.

Assumption 3. The first life-form relied on the chemical energy of minerals, not the Sun. Sunlight and lightning are both too violent, and they are uncharacteristic of the energy sources most cells use today. Moreover, photosynthesis is an immensely complex sequence of chemical reactions, requiring numerous proteins and other specialized molecules. Surface reactions on minerals, by contrast, are simple and similar to the synthesis strategies central to many cellular processes today.

Assumption 4. Metabolism came first. Many researchers claim that life requires encapsulation and membranes, but simple membranes are ill suited to let food in or waste out. Others, seduced by discoveries of modern molecular genetics, are convinced that life began with a self-replicating genetic molecule like RNA (see Chapter 16), but even the simplest genetic molecule is vastly more complex than anything in the

Miller–Urey soup. Wächtershäuser, in sharp contrast to these more conventional ideas, assumes that life began with metabolism, which he defines as a simple cycle of chemical reactions that duplicates itself.

These key ideas are amplified by one additional assumption, common to all origin theories—that of biological continuity. Today's biochemistry, no matter how intricate and dependent on specialized catalysts, has evolved in an unbroken path from primordial geochemistry. Thus, for example, Wächtershäuser's scheme builds on the observation that many of the molecules that enable today's living cells to process energy have, at their core, a mineral-like cluster of iron or nickel and sulfur atoms.

But Wächtershäuser's theory is much more than a list of assumptions. Over 100 pages of detailed chemical reactions—testable steps that lead from simple geochemical raw materials to biology—amplify his epic proposition. "You don't mind if I brag a little," he told a reporter for *Earth* in 1998, "but something like this has never been done in the entire field."

We'll look at the details of Günter Wächtershäuser's theory in Chapter 15, but the central chemical idea is that iron and nickel sulfides, notably the common iron–sulfur minerals pyrrhotite and pyrite, served as template, catalyst, and energy source for biosynthesis. In Wächtershäuser's view, simple molecules like carbon monoxide and hydrogen react on sulfide surfaces to produce larger molecules. These molecules tend to have negative charges and so they stick to the positively charged sulfide surfaces, where additional reactions build larger and larger molecules. Various surface-bound molecules begin to feed off each other, eventually forming a chemical cycle that copies itself. Voilà! It's alive! In Wächtershäuser's view, this emergent process is both inevitable and fast. "It takes maybe two weeks," he estimated in answer to a question at a Carnegie seminar.

Wächtershäuser's theory has proven more than a little controversial, and it has evoked much discussion and comment, both pro and con. "There's probably nothing there because, otherwise, people would have found it already," Jeffrey Bada says. In a recent paper in *Science*, he and his colleague Antonio Lazcano dismiss Wächtershäuser's originality: "It's not a new idea," they write, pointing to a 1955 paper by the Lithuanian-American microbiologst Martynas Ycas—a two-page "Note on the origin of life" in the *Proceedings of the National Academy of Sciences* that had comparatively little impact on the origins community. Whatever one thinks of Wächtershäuser's full-fledged and widely cited theory, it cannot be dismissed as a mere restatement of earlier ideas.

More than most scientists, Wächtershäuser seems to keep track of who cites his work, who offers praise, who follows up with relevant experiments. Though we had never met, he considered our Carnegie group a friend. His provocative theoretical ideas had influenced our experimental program. We had focused on the possibility that in the presence of minerals—especially iron and nickel sulfides—metabolism can proceed without protein catalysts. Many of our experiments served to test his ideas. So in January 1998, as a member of the Lab's seminar committee, I invited Wächtershäuser to give a lecture. He responded by phone less than a week later, and we agreed on a visit near the end of March.

Intense and earnest, combative and to the point, Wächtershäuser delivered his lecture like a legal summary to a packed, attentive room. An hour proved not nearly enough to touch on the philosophy and outline of his epic hypothesis, but many of us had read his papers and followed as one does a favorite, oft-told tale. When he did touch on other theories or dissenting views, he dismissed them quickly, efficiently, as if he were brushing away a pesky fly.

I should be clear: None of us was a Wächtershäuser disciple, nor did any of us believe everything he proposed; indeed, the idea that prebiotic chemistry and the primordial soup played no role in life's origins runs counter to the assumptions of many of our hydrothermal synthesis experiments. But his origin-of-life model is amazingly rich and detailed. Virtually every chemical step is testable by experiments. Nothing is fuzzy or left to speculation. His daring lies in the prediction of specific chemicals that arise by specific reaction pathways. Right or wrong, Wächtershäuser has produced a synthesis that will be studied by scientists for decades to come (and historians of science long after that).

Following the seminar, several of us were eager to help test the model in any way we could. That afternoon we showed him our extensive experimental facilities, described our origins research, and raised the possibility of collaboration. Our offer was for him to propose the experiments and be the lead author on any resulting publications. A week later I wrote, "We would like to explore the possibility of collaborative experimental work, in particular the effects of modest pressure on the rates of reactions you propose." He received our offers politely, but was clearly reluctant and declined to accept our help. I was left with the strong impression that Günter Wächtershäuser is eager to maintain control over both his theory and its experimental verification.

EXPERIMENTS

The hypothesis that minerals might enhance the stability of some biomolecules received a boost from a series of experiments in 1999–2000 conducted by my Carnegie colleague Jay Brandes. Brandes had arrived at the Geophysical Lab fresh from his 1996 University of Washington PhD in oceanography—a study of nitrogen chemistry in the Central Amazonian Basin of Brazil. Jay's original proposal for work at the Lab was to elaborate on the nitrogen work, but he soon shifted his efforts to our origin-of-life program. It proved a good decision.

His first paper at Carnegie, "Abiotic nitrogen reduction on the early Earth," appeared in the September 24, 1998, issue of *Nature* and immediately gained widespread attention. Brandes demonstrated that hydrothermal vents could have provided a reliable source of ammonia—an essential nitrogen-bearing compound for prebiotic synthesis. He followed up by demonstrating the synthesis of amino acids under ammonia-rich vent conditions.

In spite of these successes, studies by Stanley Miller's group on the rapid breakdown of amino acids had raised a question that would not go away. Miller and his colleagues objected to the hydrothermal-origin hypothesis, in part, because amino acids decompose rapidly at the elevated temperatures of ocean vents. How could we be making appreciable quantities of amino acids in our runs at 200°C, if these chemicals aren't stable? So Jay's next round of experiments focused on the breakdown of amino acids under a variety of geochemically relevant conditions.

Brandes's new experiments found evidence that the amino acid leucine, which breaks down in a few minutes in 200°C pressurized water, may persist for days when pyrrhotite, an iron–sulfur mineral commonly found at submarine volcanic vents, is added to the mix. While the exact protection mechanism is still under study, Brandes's experiments seemed to demonstrate that minerals may greatly enhance the stability of essential biomolecules.

Two other intriguing lines of research point to the complexity of determining amino acid stability. One recent set of experiments focuses on fossil bones, which often preserve ancient proteins. Fossil bones are composite materials in which a strong but brittle mineral matrix interweaves with flexible fibers of the proteins osteocalcin and collagen. Unprotected, these proteins would break down in a matter of a few centuries, but some fossil bones are known to preserve proteins for many millions of years. Recent work in Andrew Steele's laboratory has even revealed hints of fossil collagen in dinosaur bones more than 70 million years old. The secret to such exceptional preservation is strong bonding between minerals and the amino acid constituents of the proteins. Minerals can thus protect and preserve amino acids.

Additional evidence for amino acid survivability comes from experiments conducted by Stanford graduate student Kono Lemke and geochemist David Ross of the U.S. Geological Survey laboratories in Menlo Park. Working with financial support from our NASA Astrobiology Institute grant, they placed amino acids in a superhot pressure cooker and watched for them to decompose. But, unlike the protocol of our restrictive gold-tube experiments, Lemke and Ross employed an exotic flexible "gold bag" apparatus—a great improvement on the sealed-tube experiments.

The heart of the gold-bag apparatus is a thin-walled flexible bag about the size of a grapefruit, meticulously crafted from pure gold foil. The bag opens at one end into a titanium-valve system, with which Lemke and Ross filled and emptied their experimental solutions. They immersed the entire gold assembly in a water-filled pressure chamber that was compressed to several hundred atmospheres and heated to several hundred degrees—conditions similar to those found at deep-ocean hydrothermal zones. Once they loaded and sealed the apparatus, hot, pressurized water compressed and heated their sample container uniformly on all sides.

The principle of the gold-bag apparatus is much the same as that of our smaller gold-tube experiments. Both rely on soft, inert, deformable gold to exert uniform pressure and temperature on a fluid sample. The great advantage of the gold-bag setup is that small samples of the reacting fluid can be extracted every few hours or days throughout the duration of a long experiment. But trade-offs are a fact of scientific life. The disadvantage is that the thin-walled gold bag is frustratingly fragile and can be a pain to use. A slight miscalculation and the bag will rip, ruining an experiment and requiring a tedious and exacting welded repair. But once properly filled with reactants, the device usually works wonderfully well.

Lemke and Ross gingerly placed the bag into a water-filled pressure vessel, slowly filled the gold container with a solution of the amino acid glycine and water, sealed the assembly, and ramped up the pressure and temperature. They ran their samples for weeks, monitoring the solution, watching for the glycine to decompose. Over time, the concentration of glycine steadily declined, but even though they didn't use minerals in their experiments, they observed a much slower rate of breakdown than had been reported in previous studies at lower pressure.

In addition, they found a surprisingly fast rate of peptide-bond formation—amino acids linking together to form molecular chains. This result was unexpected, because hot water tends to break apart rather than form peptide bonds. (That's one reason boiled foods are so squishy—the sturdy bonds between amino acids that give food texture break down.) Once formed, these peptides decomposed rather quickly, but their formation pointed to more complex behavior than had been expected.

Lemke and Ross found hints of another potentially important behavior in their gold bag. Single amino acid molecules and small clusters with just two or three molecules linked by peptide bonds readily dissolve in water at room temperature, but longer peptide chains proved much less soluble. Lemke and Ross imagine a scenario in which peptides form rapidly in vents and are then exposed to the cooler seawater. Given a high enough concentration of long amino acid chains, these molecules might separate out as a relatively stable concentrated phase—just the kind of emergent molecular selection and organization that life's origin required. There's a lot more work to be done, but it appears that the book is not yet closed on amino acid stability in hydrothermal systems.

FIXING CARBON

The most fundamental biological reaction—and one of our group's primary goals in prebiotic-synthesis experiments—is carbon fixation, the incorporation of more carbon atoms (starting with carbon dioxide) into organic molecules. After all, the first chemical step in the path to life must be to make bigger molecules, like amino acids and sugars, out of smaller ones, like carbon dioxide, ammonia, and water. Such reactions occur rapidly in our experiments, but they follow two rather different paths, depending on the mineral employed.

Many common minerals, including most oxides and sulfides of iron, copper, and zinc, promote carbon addition by a routine, industrially important process known as the Fischer–Tropsch (F–T) synthesis. In its idealized form, F–T synthesis builds long chainlike organic molecules from carbon dioxide and hydrogen that are exposed to hot, dry metal surfaces. Our gold-tube experiments and studies in several other labs display similar reactions in the presence of *wet* mineral surfaces at high pressure, though a lot less efficiently than the industrial process.

Field studies complement these experiments. Recent intriguing analyses of organic molecules emanating from hydrothermal vents reveal similar Fischer–Tropsch-like products, and it now appears that F-T synthesis constantly manufactures larger organic molecules from smaller building blocks in Earth's hydrothermal zones. Many of these molecules are hydrocarbons of the type that form petroleum. (Who knows, maybe Tommy Gold is correct and at least some petroleum forms abiotically at depth.)

Alternatively, when we use nickel or cobalt sulfides, we observe that carbon addition occurs primarily by the insertion of carbon monoxide, a molecule with one carbon atom and one oxygen atom, which readily attaches itself to nickel or cobalt atoms. By repeating these simple kinds of reactions—add a carbon atom here, an oxygen or hydrogen atom there—over and over again, new and more complex organic molecules emerge.

One conclusion seems certain: Mineral-rich hydrothermal systems contributed to the early Earth's varied inventory of potential bio-building blocks. With tens of thousands of miles of deep-ocean hydrothermal ridges, billions of cubic kilometers of warm wet crust, and hundreds of millions of years to process the raw materials, organic molecules must have been produced in prodigious amounts. But the Geophysical Lab synthesis experiments have done more than simply add to the catalog of interesting molecules that could have been formed on the early Earth. These experiments are now uncovering something quite new about the possible role of minerals in the origin of life.

Previous origin-of-life studies, such as those of Günter Wächtershäuser, treat minerals essentially as solid and relatively stable platforms for synthesis and assembly of organic molecules. But our experiments reveal another, more complex behavior that may have important consequences for origin-of-life chemistry. We find that in the presence of high-temperature and high-pressure water, minerals often start to dissolve. In the process, the dissolved atoms and molecules from the minerals themselves become crucial reactants in the prebiotic milieu. Sulfur, dissolved from sulfide minerals, combines with carbon dioxide and water to form thiols and thioesters—reactive molecules that can jump-start new synthetic pathways.

Even more dramatic is the behavior of iron, which can dissolve in water to form brilliantly colorful organic solutions. After one experimental run, George Cody bounced from office to office on the second floor, showing off a particularly striking orange-red solution he had just extracted from a pressure capsule. The deep color was exciting because it pointed to the formation of iron complexes—iron atoms surrounded by a starburst of organic molecules. Chemists have long known that similar iron complexes promote chemical reactions, so Cody speculated that our cheerful solutions might contain a kind of primitive catalyst that promoted the assembly of more complex molecular structures.

Such behavior is not entirely unexpected, for hydrothermal fluids are well known to dissolve and concentrate mineral matter. Many of the world's richest ore deposits arise from hydrothermal processes. Similarly, spectacular sulfide pillars tens of meters tall grow rapidly at volcanic vents called black smokers, where rising plumes of hot, mineral-rich solutions contact the frigid water of the deep ocean.

Yet there's so much we don't know about hydrothermal systems and the chemical processes that might occur in their vicinity. And in spite of their prevalence, the role of these dissolved ingredients has not yet figured significantly in origin scenarios. No one yet knows how this rich mix of organic compounds and dissolved minerals might influence the synthesis and assembly of biomolecules. But we're poised to find out.

9

Productive Environments

*The limits of life on this planet have expanded to such a
degree that our thoughts of both past and future life have been
altered.*

Kenneth Nealson, 1997

E ven as the debate between Miller's advocates and the ventists was
heating up, an explosion of new research dramatically changed the
research community's view of the emergence of biomolecules. When
Miller first reported organic synthesis on a benchtop in 1953, the results seemed almost magical. Fifty years ago, no one could have predicted how easy it would be to make amino acids, sugars, and other key
biomolecules from water and gas. But the more scientists study carbon
chemistry in a wide range of plausible, energetic prebiotic environments, the more diverse and facile organic synthesis seems to be. It
now appears that anywhere energy and simple carbon-rich molecules
are found together, a suite of interesting organic molecules is sure to
emerge. It's all a matter of environment, and it now appears that the
universe boasts an extraordinary range of productive environments.

MOLECULES FROM DEEP SPACE

The last place you might think to look for life-forming molecules is the
black void of interstellar space, but new research reveals that organic
molecules from space must have predated Earth by billions of years.
Deep space, we now realize, is home to immense tenuous clouds where
carbon, hydrogen, oxygen, and nitrogen combine in complex sequences
of reactions.

A research team at NASA Ames Research Center at Moffett Field,

California, led by veteran astrochemist Louis Allamandola, has simulated the ultracold deep-space environments of these so-called dense molecular clouds (though these vast volumes of dust and gas are far less dense than the highest vacuum attainable on Earth). A typical interstellar cloud harbors only a measly million atoms per cubic inch, at temperatures colder than $-100°C$. Such high vacuums and frigid temperatures would seem to preclude any sort of chemical reaction, but in these remote regions, minute ice-covered dust particles are subjected to ultraviolet radiation from distant stars. Gradually, as molecules absorb this radiation, they become sufficiently reactive to form larger collections of atoms. Radio astronomers have long recognized the distinctive signatures of numerous organic species in these clouds. Each type of molecule absorbs or emits characteristic wavelengths of light—features that appear as sharp lines on a radio spectrum. The most abundant molecules are the diatomic and triatomic species, such as CO, H_2, CO_2, and H_2O, but more than 140 different compounds are known, including many larger molecules with a dozen atoms or more.

Theorists easily explain such molecular diversity. They calculate the efficiency with which small cold molecules condense onto tiny dust particles, forming submicroscopic ice coatings. They predict details of how icy particles occasionally absorb ultraviolet radiation, which can shuffle electrons and trigger chemical reactions. They plot reaction cascades by which small groups of atoms clump together and slowly cause new larger molecules to accumulate in the cloud. Eventually, under the pervasive inward pull of gravity, local regions of a molecular cloud can collapse into a new planetary system with a central massive star and an array of planets and moons. As each body forms, a steady rain of organic-rich comets and asteroids contributes to the life-forming inventory. So, the theorists tell us, organic molecules inevitably constitute part of any planet-forming mix.

Regardless of how convincing a theory may sound, experiments carry a lot of weight in science. Allamandola and co-workers' experiments at NASA Ames have exploited an elegant chilled vacuum chamber, about 8 inches in diameter, crafted of shiny stainless steel, and equipped with thick glass observation ports, to produce suites of organic molecules. First, they introduce a fine spray of simple gas molecules, such as water, carbon monoxide, methane, and ammonia, into the chamber, where the gases freeze onto an aluminum disk. Then they bathe the thin ice layer with a beam of ultraviolet radiation, which triggers the formation of larger molecules—compounds that match the distinctive molecular emissions from those distant clouds. [Plate 5]

The NASA team has used their benchtop apparatus to produce a rich variety of interesting molecules: reactive nitriles, ethers, and alcohols abound, as do ringlike hydrocarbons. One set of experiments yielded nitrogen-bearing precursor molecules to amino acids. Another set generated long chainlike molecules reminiscent of the building blocks of cell membranes.

Evidence from space amply buttresses these nifty experiments. The Murchison meteorite and many other carbon-rich meteorites are loaded with organic molecules thought to be of extraterrestrial origin. Comets, too, are known to be rich in the molecular precursors of life, as are the microscopic interplanetary dust particles that incessantly drift down to Earth's surface. Armed with their vacuum chamber, the Ames team can reproduce the supposed deep-space synthesis processes in the lab. Theory, observations, and experiments agree: The prebiotic Earth was seeded abundantly with extraterrestrial organics.

Nevertheless, the Miller crowd is unpersuaded by these studies, too. Says Miller, "Organics from outer space, that's garbage, it really is." Jeff Bada echoes, "Even if cosmic debris struck the prebiotic Earth at 10,000 times the present levels, the resultant prebiotic soup would still have been much too weak to engender life."

MOLECULES FROM GIANT IMPACTS

Meteorites and comets carry a rich inventory of organic molecules, but can these molecules survive the catastrophic insults of collisions with Earth? Deep-space synthesis, no matter how fecund, would be irrelevant to life's origin if the intense temperatures and pressures of impact disintegrated molecules.

It's hard to imagine an environment more destructive to life and its molecules than the shattering surface impact of an asteroid or comet. Nevertheless, carbon-rich meteorites like the Murchison contain a significant store of amino acids and other potential biomolecules; evidently impacts don't destroy all organic molecules. In fact, recent experiments suggest quite the opposite. Jennifer Blank and her colleagues at the Lawrence Berkeley National Laboratory, in Berkeley, California, use a giant experimental gas gun that hurls hyperfast chunks of metal at innocent rocks. Their goal is to trace the fates of organic molecules during these violent collisions.

Blank's experiments begin with a flattened cylindrical stainless-steel capsule about 1 inch in diameter that is filled with a solution of five different amino acids in water. She carefully positions the sealed sample in a metal well—the target at the end of a 40-foot-long gun barrel.

"Clear the room!" she demands, as they close the gun chamber.

A technician powers up her weapon. "Three, two, one, fire!" and *blam!*, a tremendous shock wave shakes the building as a massive metal projectile hurls down the barrel at more than 4,000 miles per hour and squashes her neatly prepared sample like a bug. For a few microseconds, the amino acid solution experiences pressures in excess of 200,000 atmospheres at temperatures approaching 900°C.

Then the fun begins. Blank pries out her deformed steel capsule and mills down the metal to extract a few drops of liquid. The original clear solution has turned a dark brown color—something interesting has happened to the amino acids. The organic chemists' standard analytical techniques, chromatography and mass spectrometry, tell the story. To be sure, most of the original amino acids are lost in every run. But, remarkably, some of the delicate molecules react with each other to form pairs of amino acids. The formation of these peptide bonds between amino acids is a crucial step in the assembly of proteins.

Jennifer Blank's highly publicized conclusion: Impacts on the early Earth may have reduced the quantity of organic molecules, but at the same time they increased the diversity of complex prebiotic chemical species.

MOLECULES FROM HOT ROCKS

Of all scenarios for the prebiotic production of organic molecules, none is more original (and correspondingly controversial) than the idea of Friedemann Freund, a longtime researcher at the NASA Ames Research Center. He claims that igneous rocks were, and still are, a principal source of Earth's organic molecules. "Maybe," he remarked to Wes Huntress, the Geophysical Lab's director, "the next chapter in the origin of life is written in the solid state—in the dense, hard, seemingly hostile matrix of crystals."

Freund, who is as persistent and unflappable as anyone you're ever likely to meet, smiles and quietly presents his case. Tall, lean, with a shock of graying hair, he speaks gently, with a slight German accent and lots of eye contact. He's always ready to talk about what he's doing and seldom expresses the slightest doubt that he's onto something important.

Here's how he claims it happens. At high temperatures, every melt contains some dissolved impurities. Molten rocks are no exception; they always incorporate a little bit of water, carbon dioxide, and nitrogen. As the melt cools, minerals begin to crystallize one after another. The first mineral might be rich in magnesium, silicon, and oxygen, but inevitably it will also incorporate a small amount of carbon and nitrogen—elements that don't easily enter the crystal lattice. These residual elements concentrate along crystal defects—zipperlike elongated spaces where the foreign atoms can react and, according to Freund, ultimately form chainlike molecules with a carbon backbone. Freund suspects that every igneous rock has the potential to manufacture such organic molecules. When the rocks weather away, so the story goes, they release vast amounts of organic carbon into the environment.

Many scientists would say that's a wacky idea. "I am a hundred percent sure that the Freund paper is utter nonsense," asserts Washington University mineralogist Anne Hofmeister. "Most igneous rocks form from an incandescent melt at temperatures greater than 1,000°C—temperatures at which even the hardiest organic molecule is fragmented into carbon dioxide and water. By contrast, organic contamination is everywhere in our environment." What causes Freund's observed organics? "It's surface residues," Hofmeister says, "probably sorbed out of the air."

Freund rests his case on two sets of samples he has been studying for almost a quarter century. Two-inch-long synthetic magnesium oxide (MgO) crystals, produced by cooling a white-hot MgO melt from 2,860°C, serve as a simple model system. Pure MgO should be clear and colorless, but Freund's crystals have a cloudy, turbid interior, suggestive of pervasive impurities. Infrared spectra reveal the sharp absorption features of carbon-hydrogen and oxygen-hydrogen bonding, both characteristic of organic molecules. Studies of the crystals' unusually high electrical conductivity and other anomalous properties have further convinced him that the supposed MgO crystals are loaded with excess carbon and hydrogen. The clincher: Subsequent analyses of molecules extracted from crushed MgO crystals reveal the presence of carboxylic acids, which just happen to be essential molecules in the metabolism of all cells.

Studies of natural gem-quality olivine, an attractive green mineral that is among the commonest constituents of igneous rocks, complement Freund's work on synthetic MgO. Once again, his spectroscopic studies revealed C–H and O–H bonds; once again, he extracted organic molecules from crushed powders. Olivine crystals hold an astonishing 100 parts per million carbon, he claims. Furthermore, much of that carbon occurs in biologically interesting, chainlike organic molecules.

Others remain unconvinced. Caltech mineralogist George Rossman duplicated some of Freund's olivine results with dirty crystals. "I ran a sample of ours that had been standing around for a while," he told Anne Hofmeister in 2002. "It had the organic bands. I washed it off with organic solvent and re-ran it. No organic bands." Organic contamination is everywhere, so any surface—especially any powder—no matter how well cleaned, will quickly become loaded with adsorbed organic molecules. Freund counters that the types of molecules he extracts, carboxylic acids, are not typical of any ordinary environmental contamination. They must have come from inside the mineral.

Freund had won relatively few converts by the summer of 2003, when he came to George Cody's lab to duplicate his extraction of molecules from olivine. For several weeks, a white-coated Freund was an amiable fixture at Cody's lab bench. He meticulously washed and powdered the semiprecious stones, extracted carbon compounds with strong solvents, and analyzed the samples with Cody's battery of high-tech instruments. Sure enough, every crystal seemed to release a small hoard of carbon-rich molecules. There wasn't much, certainly, but the volume of igneous rock that has formed and eroded over the course of geological history is immense. So, by Freund's estimates, solid rocks have provided one of Earth's largest and most continuous sources for the emergence of biomolecules.

Scientific progress involves a long process of hypothesis and testing, bold claims, and critical counterarguments. Not surprisingly, Freund's hypothesis has received a lot of scrutiny and not a little disdain. But those unexplained carboxylic acids can't be ignored. And so, for the time being, the jury is still out.

THE MULTIPLE-SOURCE HYPOTHESIS

Where did life's crucial molecules form? In spite of the polarizing advocacy of one favored environment or another by this group and that, experiments increasingly point to the possibility that there was no single dominant source.

It's not a matter of Millerites versus ventists, or deep space versus Earth's surface. Many ancient environments boasted carbon atoms and sufficient energy to initiate their chemical transformations. Many environments must have contributed to the prebiotic inventory. Lightning-sparked gases were a major source, to be sure, as were UV-triggered reactions high in the atmosphere. Deep in the ocean, in environments ranging from lukewarm to boiling hot, molecules must have been made in abundance, as they certainly were within some reactive rocks of the crust (and, if Tom Gold is correct, perhaps in the much deeper mantle). A wealth of organic products also rained down from space, formed in remote dense molecular clouds and concentrated in the carbon-rich meteorites and asteroids that coalesced to make our planet.

The bottom line is that the prebiotic Earth had an embarrassment of organic riches derived from many likely sources. Carbon rich molecules emerge from every conceivable environment. Amino acids, sugars, hydrocarbons, bases—all the key molecular species are there.

So the real challenge turns out to be not so much the making of molecules, but the selection of just the right ones and their assembly into the useful structures we call macromolecules. That process required a higher level of emergence.

Interlude—Mythos Versus Logos

People of the past . . . evolved two ways of thinking,
speaking, and acquiring knowledge, which scholars
have called mythos and logos. Both were essential.
Karen Armstrong, *The Battle for God,* 2000

"Whoa, wait a minute!" My wife, Margee, sets aside my draft, her expression a cross between confusion and exasperation. "Is any of this stuff true? Who's to say you're not just writing another creation myth?"

"What do you mean? This is science." How could she miss the point? "There's a big difference between myth and science!"

"But you're just making up a story. It's *really* ancient history—no one will ever know for sure how life started." She's warming up to the debate. "Besides, you're constantly saying 'We don't know' and 'The jury's still out.' Can you be sure about *anything*?"

"Gimme a break!" was about the cleverest comeback I could think of, as I turned back to the word processor.

So which is this book? *Logos*? *Mythos*? Some combination of the two? Am I writing the truth, or only just-so stories?

The distinction is not always clear-cut. The studies of life's origin are in some ways like the efforts of archaeologists to document the history of ancient Troy. Troy fell to the Greeks in about 1190 BCE and eventually was buried under the litter of later cities. Real people were born, led their lives, and died in Troy; nevertheless, most of that rich, poignant history is lost forever. We learn fragments of the truth from excavations, artifacts, and ancient documents. But mythology always lurks in the background. The *Iliad,* the *Odyssey,* and the *Aeneid* inevitably color our understanding of the great city's past.

Life, too, emerged through some real process. Molecules formed, they combined, they began to replicate. Much of that history is also lost forever. We will never know exactly where or when the first living entity arose, nor is it likely that every chemical detail of the process will ever be known for certain. Scientists flesh out the process with their own favorite origin stories: Miller's primordial soup, Gold's deep hot biosphere, Wächtershäuser's sulfide surfaces. We tend to favor the stories told by our friends or our mentors, while discounting those of our rivals. And even if we do succeed in making life in the lab, there's no guarantee that that's exactly the way it happened 4 billion years ago.

Nevertheless, science and myth differ in a fundamental way. Scientific stories must win support through logically sound theory, rigorously reproducible lab experiments, and independently verifiable observations of nature. A scientific hypothesis must make unambiguous predictions. If those predictions fail, the story is deemed false by the scientific community and is cast aside. Today we may debate the details of the process, but all scientists agree that there must be a true origin-of-life story. That truth is our common goal.

There's so much we don't know and, as you have undoubtedly noticed, much of this book is qualified with phrases of uncertainty. Hardly an experiment or theory goes unchallenged, and groups of researchers often reach diametrically opposed conclusions. But we have attained a vibrant stage in origins research, one in which we are increasingly aware of what we don't know and, consequently, are increasingly focused on what we must learn. A sustained, confident international program of research has supplanted the naïve optimism of the 1950s and 1960s. And so the scientific stories come thick and fast as theory, experiment, and observation winnow the universe of possibilities.

Part III

The Emergence of Macromolecules

The beginning and end points of life's emergence on Earth seem reasonably well established. At the beginning, more than 4 billion years ago, life's simplest molecular building blocks—amino acids, sugars, hydrocarbons, and more—emerged inexorably through facile chemical reactions in numerous prebiotic environments, from deep space to the deep crust. A half-century of compelling synthesis research has amplified Stanley Miller's breakthrough experiments. Potential biomolecules must have littered the ancient Earth.

The end point of life's chemical origin was the emergence of the simple, encapsulated precursors to modern microbial life, with all of life's essential traits: the ability to grow, to reproduce, and to evolve. Top-down studies of the fossil record hint that such cellular life was firmly established almost 4 billion years ago.

The great mystery of life's origin lies in the huge gap between molecules and cells. Ancient Earth boasted oceans of promising biomolecules but, like a pile of bricks and lumber at a building supplier, more than a little assembly was required to achieve a useful structure. Life requires the organization of just the right combination of small molecules into much larger collections—macromolecules with specific functions. Making macromolecules from lots of little molecules may sound straightforward, but what most books don't mention is that for every potentially useful small molecule in that prebiotic environment, there were dozens of other molecular species with no obvious role in modern biology. Life as we know it is incredibly picky about its building blocks; the vast majority of carbon-based molecules synthesized in prebiotic processes have no obvious biological use whatsoever. That's why, in laboratories around the world, many origins researchers have shifted their focus to the emergent steps by which just the right molecules might have been selected, concentrated, and organized into the essential structures of life.

10

The Macromolecules of Life

To purify and characterize thoroughly all [biomolecules]
would be an insuperable task were it not for the fact that
each class of macromolecules . . . is composed of a small,
common set of monomeric units.
                    Lehninger et al., *Principles of Biochemistry,* 1993

We are chemical beings. Every living organism, from the simplest microbes to multicellular fungi, plants, and animals, incorporates thousands of intricate molecular components. All of nature's diverse life-forms grow, develop, reproduce, and respond to changes in their external environment—vital tasks that must be accomplished by exquisitely balanced cascades of chemical reactions.

The more biologists learn about life, even the most "primitive" single-celled organisms, the more amazingly complex life seems to be. Everywhere you look, living entities have found their niche, and they survive in wonderfully varied ways. Indeed, in a sense, chemical complexity seems synonymous with life. Yet emergent systems, however complex, are usually built from relatively simple parts, and life is no exception.

One of the transforming discoveries of biology is that all known life-forms rely on only a few basic types of chemical reactions, and these reactions produce a mere handful of molecular building blocks. Virtually all of life's essential construction materials are carbon-based organic molecules that combine by the thousands to form layered enclosures or chainlike polymers. In every instance, just a few kinds of small molecules assemble into a great variety of larger structures.

In the early nineteenth century, conventional wisdom held that life's chemical compounds formed by their own mysterious rules, perhaps governed by a "vital force." Many scholars assumed that the nascent science of chemistry applied only to the inorganic world—the world of rocks, minerals, and metals. This perception changed in 1828, when the young German chemist, mineral collector, and gynecologist Friedrich Wöhler demonstrated that biological molecules are no different in principle from other chemicals. He combined the common laboratory reagent cyanic acid with ammonia and succeeded in producing urea, which is extracted in the kidneys and found in urine. Wöhler's letter announcing his important discovery displayed a sense of whimsy not always associated with German academicians: "I can no longer, as it were, hold back my chemical urine: and I have to let out that I can make urea without needing a kidney, whether of man or dog." By employing straightforward lab techniques to produce a chemical substance known only from life, Wöhler convinced his colleagues of the ordinariness of organic chemistry.

MODULAR LIVING

Perhaps the most distinctive characteristic of the molecules of living organisms is their modular design. This familiar strategy is similar to that of modern architects, who rely for the most part on standard building materials. They use mass-produced bricks, beams, windows, doors, stairs, lighting fixtures, and so on to assemble an almost infinite variety of commercial and private buildings.

You don't have to build in that way. A multimillionaire acquaintance recently created the most extraordinary mansion, with every square foot of walls and floor, every light and bath fixture, every door and window custom-designed and hand-crafted. It's an amazing house, with secret passages, unexpected nooks, and hidden closets, all lovingly constructed of the finest woods, stone, and other extravagant materials. Such personalized craftsmanship is wonderful to see, but inordinately expensive. Most of us choose a more economical path. By relying on a few standard construction modules, buildings are faster and cheaper to design and build. But modularity doesn't imply uniformity. You can design a unique dwelling of almost any size or shape from simple components available through any hardware or building supply store.

The same modular principle holds for life's carbon-based molecular building blocks. Four key types of molecules—sugars, amino acids, nucleic acids, and hydrocarbons—exemplify life's chemical parsimony.

Sugars are the basic building blocks of carbohydrates, Earth's most abundant biomolecules. Many common sugars in our diets, including the fructose of honey, the sucrose of cane sugar, and the glucose of fruits, are small molecules consisting of at most a few dozen atoms. These energy-rich molecules incorporate carbon, oxygen, and hydrogen typically in about a 1:1:2 ratio. But most of life's sugar molecules are locked into macromolecules—countless individual sugar molecules linked together to form giant polymers, such as the fibrous cellulose of plant stems or the bulky starch of potatoes.

Amino acids are small molecules that link together to form proteins. Proteins serve as the chemical workhorses of life, with myriad vital functions: They form tissues and strengthen bone; they act as hormones to control glandular functions; they clot blood, digest food, and promote the thousands of chemical operations essential to life. All proteins form from long chains of hundreds to thousands of individual amino acid molecules, lined up like beads on a string. These chains fold into the most wonderful shapes, each protein folded in such a way as to accomplish a specific chemical task.

The vital genetic molecules DNA and RNA are also lengthy chainlike polymers, assembled from small molecules called nucleotides. Each nucleotide, in turn, is constructed from three small molecular parts: a 5-carbon sugar (ribose in RNA, its cousin deoxyribose in DNA); a base (one of five closely related ring-shaped molecules), and a phosphate group (a tiny cluster made up of a phosphorus atom surrounded by four oxygen atoms). RNA consists of a long chain of single nucleotides, while in DNA two such chains link together and twist into the famed "double helix" structure.

Finally (as we'll see in the next chapter), arrays of elongated hydrocarbon molecules called lipids, including a wide variety of fats and oils, coalesce in every cell to form membranes, store energy, and perform other critical functions.

The most enticing aspect of Stanley Miller's experiment and the discoveries of subsequent prebiotic researchers was that they synthesized components (or at least close relatives) of all four groups—carbohydrates, proteins, nucleic acids, and lipid membranes. No wonder so many researchers were optimistic that an understanding of life's emergence was at hand.

But this molecular catalog of success ignores a puzzling part of the prebiotic synthesis story. For every useful molecule produced, many other species with no obvious biological function complicate the picture. Take sugar molecules, for example. All living cells rely on two kinds of 5-carbon sugar molecules: ribose and deoxyribose. Sure enough, several plausible prebiotic synthesis pathways yield small amounts of these essential sugars. But for every one of these molecules produced, many other 5-carbon sugar species also appear—xylose, arabinose, and lyxose, for example. Adding to this chemical complexity is a bewildering array of more than 100 3-, 4-, 6-, and 7-carbon sugars, in chain, branching, and ring structures—of which life uses only a small handful.

As if that weren't enough of a problem, life is even choosier about its molecules. Many organic molecules, including ribose and deoxyribose, come in mirror-image pairs. These left- and right-handed varieties are in most respects chemically identical. They possess the same chemical formula and many of the same physical properties—identical melting and boiling points, for example. But they differ in their shapes, just like your left and right hands. Laboratory synthesis usually yields equal amounts of left- and right-handed sugars, but finicky life employs only the right-handed sugars, never the left.

Given the disparity between the rich variety of prebiotic molecules and the apparent paucity of biomolecules, is it possible that we're fooling ourselves? Might the earliest life-forms have used a more diverse suite of organic molecules and a different repertoire of biochemical pathways? It turns out that living cells hold clues that are now being teased out by the remarkable field of molecular phylogeny.

MOLECULAR PHYLOGENY AND
THE "LAST COMMON ANCESTOR"

The complete chemical arsenal of each living species is recorded in its unique genome, an encoded sequence of millions to billions of DNA "letters," the base pairs that form the rungs of DNA's double helical ladder. The four-letter DNA alphabet—A, G, T, and C (for the purines adenine and guanine and the pyrimidines thymine and cytosine)—is sufficient to spell out all the genetic information of any organism. What's more, all cells share the same mechanism for converting genetic instructions into the proteins that serve in many chemical and structural roles. That's why genetic engineers can use a simple bacterium, such as *E. coli*, to synthesize human growth hormone, or insulin, or other valuable pharmaceuticals.

A central assumption of molecular phylogeny is that the genomes we see today evolved over billions of years from earlier ancestral cells, via the gradual mutation of DNA sequences. Comparative examination of many genomes reveals marked similarities, as well as important differences that have inexorably arisen by this slow evolutionary process. Differences among DNA sequences suggest the evolutionary branching; the more dissimilar the sequences of two species, the longer ago they are likely to have split from a common ancestor. DNA sequences that show little variation among many diverse species (so-called highly conserved sequences) are more likely to represent ancient, essential biochemical traits.

The power and promise of phylogenetic analysis is epitomized in a remarkable recent study of early English literature. University of Cambridge biochemists Adrian Barbrook and Christopher Howe teamed with British literary scholars to analyze 58 different fifteenth-century manuscript copies of the Wife of Bath's Prologue from Geoffrey Chaucer's *The Canterbury Tales*. No copy of the late fourteenth-century original in Chaucer's hand is known, and significant variations among the many early hand-copied sources raise doubts regarding the author's original text.

For their Chaucer analysis, Barbrook, Howe, and co-workers employed the same computer techniques used by evolutionary biologists to identify the most primitive organisms. They entered each of the Prologue's 850 lines from all 58 versions into the computer and searched for textual similarities and differences. Features common to most sources presumably reflected the original text, while variations that arose from copying errors or deliberate changes were used to construct a kind of genealogy of the manuscripts.

The British team found that 44 manuscripts fell neatly into 5 groups, evidently descended from 5 different copying sources. (The remaining 14 showed more extreme deviations and were eliminated from consideration.) One group of 11 manuscripts proved crucial, for it lay much closer to the presumed original than the others. Surprisingly, these 11 texts had received comparatively little study from literary scholars. "In time, this may lead editors to produce a radically different text of *The Canterbury Tales*," Barbrook and colleagues concluded.

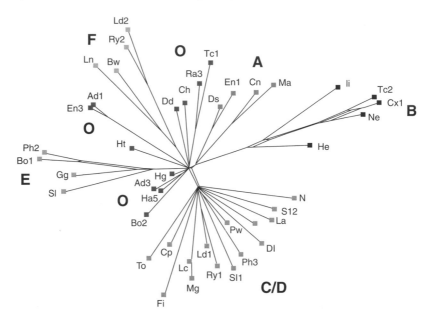

Computer-aided analysis of several dozen hand-copied manuscripts of Geoffrey Chaucer's *Canterbury Tales* points to several distinct groups of texts, denoted by letters. Each manuscript is represented as a line on this diagram; the length of the lines corresponds to the amount of deviation from a presumed primary source. Of these, the groups marked "O" appear to be closest to Chaucer's lost original (after Barbrook et al., 1998).

Molecular phylogenists adopt the same analytical approach. Their "texts" consist of strands of genetic molecules—DNA or its close cousin, RNA—with long sequences of the genetic letters. Thanks to new, rapid sequencing techniques, numerous genomes have been documented, including more than a hundred genomes of various microbes. Phylogenists use powerful computer algorithms to compare their genetic texts.

The universally acknowledged pioneer of molecular phylogeny is University of Illinois geneticist Carl Woese, a shy and retiring maverick who had to wait many years for widespread scientific acceptance of his ideas. In a landmark 1977 article in the *Proceedings of the National Academy of Sciences*, "Phylogenetic structure of the prokaryotic domain: The primary kingdoms," Woese and Illinois colleague George Fox uprooted what had become the firmly established tree of life. Prior to the "Woesian revolution," most biologists recognized five kingdoms of living organisms: animals, plants, fungi, protoctists (complex single-celled organisms with a membrane-bounded nucleus), and bacteria (metabolically diverse organisms that lack a nucleus). Based on his laborious phylogenetic analyses (which were much more difficult in the 1970s than they are today), Woese realized that animals, plants, fungi, and eukaryotes are remarkably similar in their biochemical characteristics and so constitute just one domain of life, the Eukarya. Prokaryotes, on the other hand, displayed astonishing chemical diversity and fell into two distinct kingdoms, which he called Bacteria and Archaea. Bacteria were well known for their role in causing disease, but the Archaea, that as prokaryotes resemble other bacteria in many ways, constituted an unrecognized group of microbes. So contentious was this view of life that it took almost two decades to gain general acceptance.

In retrospect, the lateness of Woese's discovery should not seem too surprising. Most species in the two microbial kingdoms, Bacteria and Archaea, occur as nondescript microscopic spheres or rods that are virtually impossible to distinguish, even with sophisticated chemical and physiological tests. Consequently, previous workers lumped them all together. Only after the application of molecular phylogeny did the profound differences become obvious.

Woese's initial intention was to unravel aspects of evolution—to establish a top-down family tree of life and infer the complex history of branchings from parent to daughter species. This evolutionary pursuit led quickly to insights regarding the nature of the last common ancestor. He proposed, for example, that Archaea and Bacteria arose long before Eukarya.

Other discoveries followed. Many of the most primitive microbes are extremophiles that thrive at elevated temperature. To many researchers, Woese's discoveries thus lent credence to the idea of a deep, hot emergence of life. Other scientists, however, quickly took issue with that conclusion. Deep microbes might well have been the only survivors following a massive asteroid impact and thus, by default, became the last common ancestors of life on Earth today. But those heat-loving microbes might well have evolved from strains of earlier, surface-dwelling cells.

Recent studies point to other intriguing possible characteristics of the presumed last common ancestor. For example, many of the most primitive Archaea are autotrophs, organisms that synthesize their own

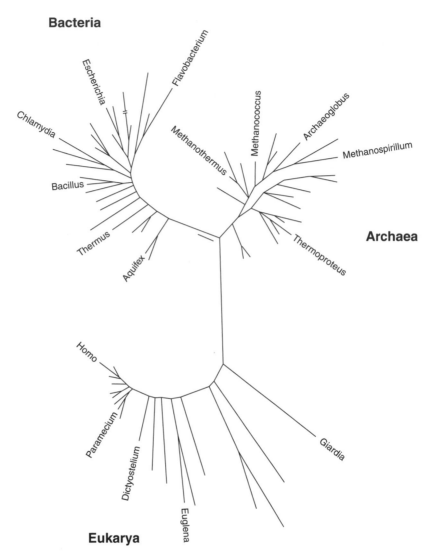

Carl Woese compared the genetic sequence of many different organisms, especially microbes, to construct a phylogenetic tree of life. He found three distinct branches of life, including the previously unrecognized Archaea. The majority of the most deeply rooted organisms are extremophiles living at high temperature (after Pace, 1997).

biomolecules. However, many origin specialists, Stanley Miller and his colleagues among them, have long argued that the first cells must have been heterotrophs, which scavenge molecules from their environment. Once again, the controversy cannot be resolved by phylogenetic studies, for deep-dwelling autotrophs that evolved from surface heterotrophs might have been the sole survivors of a globe-sterilizing event.

Additional complexity arises from the tendency of microbes to swap sections of DNA. Remarkably, throughout the history of life, cells have picked up chunks of DNA from completely different species to form new, hybrid genomes. Thus the idealized view of an unbroken branching history of DNA evolution from parent to daughter is invalid. There is no single "last common ancestor."

In spite of these uncertainties, molecular phylogeny provides invaluable information regarding the shared biochemical heritage of all cells—information that continues to inform origins research. Two observations stand out. First, all cells employ RNA to carry genetic information and assemble proteins. The earliest genetic mechanism thus points to an ancient "RNA World" (see Chapter 16). Second, at the heart of every cell's chemical machinery lies a simple metabolic strategy called the citric acid cycle, which is powered by simple chemical reactions. Any successful model of life's emergence must thus account for the origin of this cycle (see Chapter 15).

An important conclusion from these phylogenetic studies is that Earth's earliest cells were not so different from those of today. They probably relied on similar, though simpler, biochemical pathways. Those similarities provide a clear focus for experimental studies of emergent biochemistry.

The greatest challenge in understanding life's emergence lies in finding mechanisms by which just the right combination of smaller molecules was selected, concentrated, and organized into the larger macromolecular structures like RNA and in self-replicating cycles of molecules like the citric acid cycle. But regardless of how much organic stuff was made, the primordial ocean—with an estimated volume greater than 320 cubic miles—formed a hopelessly dilute soup in which it would have been all but impossible for the right combination of molecules to bump into one another and make anything useful in the chemical path to life. Complex emergent systems require a minimum concentration of interacting agents. Many scientists have therefore settled on an obvious solution: Focus on surfaces, where molecules tend to concentrate.

Interesting chemistry takes place on surfaces where two different materials meet and molecules often congregate. The surface of the ocean, where air meets water, is one such promising place. Perhaps a primordial "oil slick" concentrated organic molecules. Evaporating tidal pools, where rock and water meet and cycles of evaporation concentrate stranded chemicals, represent another appealing location for origin-of-life chemistry. Deep within the crust and in hydrothermal volcanic zones, mineral surfaces may have played a similar role, selecting, concentrating, and organizing molecules on their periodic, crystalline surfaces. Whatever the mise-en-scène, a surface seems a logical site for life's origin.

But just suppose a collection of organic molecules could organize themselves in such a way that they provided their own surface? Now *that* would be a trick worth learning!

11

Isolation

*The self-assembly process seems to defy our intuitive expecta-
tion from the laws of physics that everything on average
becomes more disordered.*

David Deamer, 2003

Water provides the universal medium for life. All known cells are mostly water on the inside, and most are surrounded by water on the outside. That aquatic lifestyle poses a problem, however, because water is one of the best solvents. You don't want your body's cells to dissolve every time you take a bath. Life had to develop an insoluble protective membrane, but what chemical to use?

Lipid molecules, which feature hydrocarbon chains (a row of carbon atoms surrounded by hydrogen atoms), provide the perfect answer to the problem. Lipids, including various fats, oils, and waxes, are characterized by their insolubility in water—oil and water don't mix. The special phospholipid molecules that form most modern cell membranes are no exception. Each of these molecules is shaped something like a tiny bobby pin, with two long hydrocarbon chains of atoms attached to a rounded end. The two exposed hydrocarbon chains are hydrophobic ("water hating"), so most of the elongated molecule is water-repellent. By contrast, the rounded end incorporates a phosphate group (phosphorus and oxygen atoms); that hydrophilic ("water loving") end attracts water. Such molecules, with both hydrophobic and hydrophilic regions, are called amphiphiles.

SELF-ORGANIZATION

When placed in water, amphiphilic lipids deal with their love-hate relationship in a remarkable way. All natural systems tend to rearrange themselves to reduce their total energy content: A tightly stretched elastic band snaps, a precariously perched boulder tumbles to the valley below, a firecracker explodes. By the same token, a solution of lipid molecules searches for a state of lower energy in which only the hydrophilic phosphate ends contact water. In the early 1960s, Alec Bangham, a biophysicist from Cambridge, England, discovered that lipids that were extracted from egg yolk and immersed in water spontaneously organized themselves into tiny spheres—structures now known as vesicles.

The energy-reducing strategy employed by the molecules that form cell membranes is nothing short of magical. Millions of individual amphiphilic molecules quickly clump together, forming a smooth, flexible double layer of lipids—a lipid bilayer. The resilient lipid bilayer provides a simple and elegant solution to the phospholipid's ambivalence toward water. All of the hydrophobic chains of atoms point toward the middle of the structure, well away from water, while the hydrophilic phosphate ends all wound up on the outside of the cell facing the wet environment or on the inside facing the water-based contents of the cell. This arrangement accomplishes the vital functions of holding the cell together while separating its inside from the outside.

Life has perfected this task of separating the inside from the outside, but could such an emergent, self-organizing process have arisen naturally in the lifeless prebiotic soup? The answer, once again, is to be found in the laboratory. Some amazing experiments have been performed by the Swiss biochemist Pier Luigi Luisi, who has spent decades studying lipid self-organization.

Not only can Luisi and co-workers form vesicles with ease, but they also demonstrate that these structures can grow, gradually incorporating new lipid molecules from solution. They've also shown that vesicles are autocatalytic—that is, they can act as templates that trigger the formation of more vesicles. And, under the proper circumstances, vesicles can even divide—a kind of self-replication.

These intriguing emergent behaviors have led Luisi to propose the "Lipid World" scenario for life's origin. In this conceptually simple model, prebiotic lipids formed abundantly on Earth and in space. Once in solution, these lipids self-organized into cell-like vesicles that captured a primitive genetic molecule, some early, simpler version of DNA or RNA. Now the Swiss team has set its sights on incorporating self-

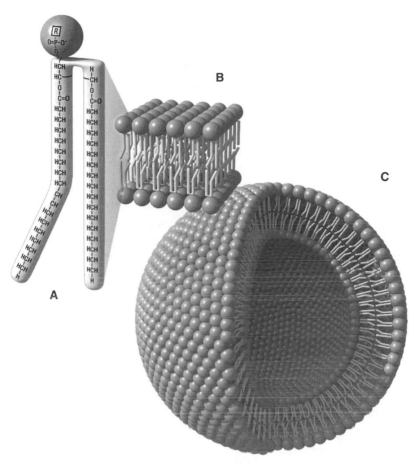

Cell membranes are formed from amphiphiles, which are elongated molecules that have both water-attracting and water-repellent ends (A). When placed in water, these molecules self-organize into a bilayer (B), which can form a spherical vesicle (C).

replicating pieces of RNA into self-replicating vesicles, perhaps even to make the first synthetic life-form. It hasn't happened yet, but the chemical pieces are close to falling into place.

MEMBRANES FROM SPACE

In the relatively brief history of origins research, a mere handful of experiments may be counted as classics. Louis Pasteur's refutation of spontaneous generation and Stanley Miller's electric-spark synthesis experiments have achieved that status, as has the novel vacuum-chamber research of Lou Allamandola and his colleagues at NASA Ames. Their results changed the way we think about life's origins. The same high regard is accorded David Deamer's remarkable discoveries of lipid self-organization in spaceborne molecules.

For almost three decades, Dave Deamer has been a popular professor of biochemistry in the University of California system. Lean, bright-eyed, with a neat graying beard and dark-rimmed glasses, he delivers lectures in a gentle and reassuring voice, like a scientific Mister Rogers. Listening to his low-key delivery, you might not guess that he is one of the world's most respected experts on the origin of life. [Plate 6]

Deamer caught the origins bug in 1975, when he took a sabbatical from the Davis campus of the University of California and went to study lipids with Alec Bangham at Cambridge. Their work revealed that the size and resilience of vesicles depends on the size and shape of the dissolved lipid molecules. In the course of these investigations, they realized that vesicles might have provided the first sheltering environment for life. If lipids existed in the early oceans, then prebiotic vesicles may have been abundant.

Deamer returned to Davis and continued this line of research, which led to his most famous experiment. That work, completed in 1988, focused on carbon-based molecules extracted from the Murchison meteorite. Ever since it landed in the Melbourne cow field in 1969, origins scientists all around the world had been bargaining and pleading for a piece of the prize. Deamer's precious 90-gram Murchison fragment, about the size of a walnut, arrived from the Field Museum in Chicago. Dave and his collaborator, chemist Richard Pashley of the Australian National University, went straight to work. Their focus was the lipids, essential biomolecules but ones that did not seem to be produced in sufficient abundance by Miller's spark process. Perhaps, they thought, carbonaceous chondrites provided those necessary raw materials for life's membranes.

Deamer and Pashley ran their sample through a series of chemical steps to break apart the dense black meteoric mass into chemically distinct fractions—steps that in some ways mimicked millennia of chemical weathering processes on the primitive Earth. Whatever molecules they found were thus likely to have occurred on the prebiotic Earth, as well. First they pulverized a portion of the meteorite into fine black powder. Their straightforward procedure involved grinding the rock while it was submerged in a liquid mixture of water, alcohol, and chloroform. These solvents don't affect the crystalline minerals that form the bulk of the Murchison, but they do dissolve different suites of interesting organic molecules. After several minutes of grinding, Deamer and Pashley poured their fine-grained slurry into a test tube, placed it into a centrifuge, and let it spin.

In the centrifuge, the pulverized meteorite solution rapidly separated into three fractions. A small pile of dense mineral fragments settled to the bottom of the tube, to be set aside just in case more studies were required. On top was a layer of water–alcohol solution, which dissolved and concentrated amino acids, sugars, and a variety of other water-soluble organic species. This fraction, too, was set aside. In the middle was a layer of chloroform, an effective solvent for any lipids the meteorite might hold. They found that the chloroform fraction had extracted more than a tenth of a percent of the meteorite fragment's mass—a surprisingly high concentration of tantalizing organic species.

Further separation was performed using chromatography. Following much the same protocols as Stanley Miller had employed decades earlier to separate his amino acids, Deamer and Pashley evaporated a portion of their chloroform sample to reveal a yellowish-brown concentrated solution. They placed a drop of this concentrate on the corner of a 4 × 4-inch glass plate that had been coated with a soft, porous white powder (an effective replacement for the older-style chromatography paper). They used ether, a colorless strong-smelling solvent, for the first chromatographic stage, stretching the dried dot into a streak. Then they rotated the plate 90 degrees and used chloroform to spread the streak into a distinctive two-dimensional array of compounds.

Viewed in daylight, the dried glass plate was unimpressive, with only the original brownish spot and a few faint yellowish areas nearby. But Deamer knew that many otherwise invisible compounds fluoresce brightly under "black light." When he darkened the room and shone an ultraviolet lamp on the plate, he was delighted to see a rich display of colors sweeping across it in a broad arc.

Deamer and Pashley identified a half-dozen distinct fluorescent regions, each with a different, as yet unknown suite of ancient cosmic organic molecules. They meticulously outlined each area by scratching the soft, powdery white surface; then they scraped off and collected powder from each of those areas into test tubes. A quick wash in chloroform was all that it took to recover the precious suites of Murchison molecules.

Keen with anticipation, the chemists placed the chemical fractions one-by-one into water and watched to see if anything interesting happened. They began with "spot 1," with molecules that had concentrated in an elongated area close to the original drop on the chromatography plate. Deamer and Pashley watched transfixed as the invisible molecules, once dispersed throughout the meteorite, spontaneously arranged themselves into tiny spheres no more than a hundredth of an inch across—about the size of many modern microbes. What's more, they found, these weren't just little drops of oil or fat floating in water. These structures had an inside and an outside. The molecules had organized themselves into bilayers, just like a cell membrane—an elegant example of emergence.

It was a breakthrough moment for origin researchers. Deamer and Pashley had shown that ancient lipid molecules, synthesized at some distant place in space and delivered intact to Earth, form tiny enclosed structures that are in many ways like the membranes encasing living cells. One of life's most basic requirements—the isolation of inside from outside—suddenly seemed to have been hard-wired into the fabric of the universe.

SELF-ORGANIZATION, REPRISE

Dave Deamer's Murchison experiments were conceptually simple and beautifully executed. So when new lipid-rich samples came along, he repeated the process.

Lou Allamandola and his NASA Ames team realized that their growing inventory of organic molecules, synthesized under simulated deep-space conditions of ultracold vacuum with ultraviolet radiation, contained a significant fraction of yellowish oily stuff just as the Murchison meteorite did. In particular, when they irradiated an ice made principally of water and alcohol with a bit of ammonia and carbon monoxide thrown in, they produced an intriguing residue of fluorescent material. Much of that material was known to be the familiar multiringed hydrocarbons known as PAHs, but other molecules appeared to have an amphiphilic character. Naturally, they turned to Dave Deamer to check it out.

Samples in hand, it took Deamer less than a day to confirm what the Ames researchers had hoped. Once the correct fraction of fluorescent molecules was concentrated, stunning vesicles appeared spontaneously in water. The press trumpeted the result, and a colorful photograph of the delicate tiny spheres graced the front page of the *Washington Post* above the headline "IN SPACE; CLUES TO THE SEEDS OF LIFE." The implications were profound: Even before the formation of planets and moons, in the tenuous vacuum of frigid space, the raw materials for life abound, ready to organize spontaneously into cell-like structures.

PYRUVATE REDUX

I got the chance to work with Dave Deamer following a conversation at one of NASA's first astrobiology meetings, in April 2000. Dave had been asked to present a keynote lecture on self-organization to the audience of geologists, chemists, biologists, and astronomers, not to mention a smattering of philosophers and ethicists.

Some scientific lecturers try to snow their audiences. Deamer is different; he meets the audience more than halfway, with comfortable metaphors, familiar examples, and elegant demonstrations. At this lecture, he held up two large beakers, both with colorless solutions. When mixed, the resulting liquid immediately became cloudy white; we were looking at the spontaneous self-organization of lipids, he explained.

At Carnegie, my group's 1996 pyruvate work had been sitting on hold for years. We knew we'd made a lot of interesting organic molecules by heating and squeezing pyruvate, but other than the fact that the reactions occurred rapidly under hydrothermal conditions, the relevance to life's origin wasn't clear. Perhaps, we thought, the yellow, oily goo that oozed out of our gold capsules held self-organizing molecules. That might be worth investigating, because we had started with a core metabolic molecule. It would be newsworthy if there were a facile path from primitive metabolism to membranes.

On hearing my story, Dave immediately invited me to his specially equipped lab at UC Santa Cruz (where he had moved in 1994) to try the experiment. The following winter, I prepared some new pyruvate-plus-water capsules and subjected each of them to two hours at 2,000 atmospheres and 250°C. I brought them, unopened, to the beautiful Santa Cruz campus.

In spite of his insanely busy schedule as faculty member in two departments, supervisor of two laboratories, and mentor to several graduate students and postdocs, Dave was a gracious and attentive host. He welcomed me to his biochemistry lab and we set to work immediately.

Once I had opened the capsules (which responded with the now familiar *bang!* and intense oily foaming), he led me step-by-step through the chloroform extraction, concentration, and preparation of a 10 × 10-inch glass plate for chromatography. I had pored over his 1989 paper several times, so it was a delight to duplicate that work with my own samples under his supervision.

Within a couple of hours, I had decorated a glass plate with a small yellow-brown dot of unknown chloroform-soluble compounds. We gently lowered the plate into a deep glass tank into which I had poured a half-inch layer of pungent ether. (The strong smell triggered a brief, vivid flashback to an early childhood moment—a menacing masked anesthesiologist bending over me, smothering my face prior to a tonsillectomy. I had to shake away the disturbing image.) As with Deamer's earlier work, the solvent pulled the glass plate's single yellow-brown spot into a long streak. Then we rotated the plate and the chloroform smeared the streak into what we hoped would be a distinctive pattern of organic compounds.

I felt more than a little tingle of anticipation as the lights went out and the UV fluorescent lamp flicked on. The results were gorgeous! A brilliant yellow, blue, and purple pattern appeared, blazing across the plate in a diffuse 7-inch-long arc of color. We were delighted to see several distinctly fluorescing areas, strikingly similar in detail to the Murchison sample [Plate 6]. Noting the correspondence, Dave suggested that we first concentrate on a blue fluorescing area most closely matching the position of his original "spot 1."

Again, we followed the 1988 procedures: Carefully mark the glass plate, scrape off the white powder from the area of interest, wash that powder with chloroform to redissolve the fluorescing molecules, and dry the extract (by this time the lab area smelled strongly of the chloroform–ether mix). Then the big test. Would my concentrated extract perform the self-organization trick?

The test was quick and easy. We applied a bit of the extract to a droplet of water on a glass slide and watched in the microscope, which used a UV light to highlight fluorescent molecules. Sure enough, tiny green fluorescing spheres appeared, like a fantastic display of Christmas lights. [Plate 7] Beautiful, but were they vesicles that trapped the surrounding liquid, or simply solid spheres? That was key to determine if we had really made cell-like bilayer membranes.

Deamer's technique was to repeat the microscope observations, but this time starting with a strongly fluorescing red dye in the water. For a second time he applied a bit of the extract to the water and, once again, green fluorescing spheres formed. If we had made hollow vesicles, then they would capture the distinctive red dye. To find out, Dave carefully flushed the slide with new, nonfluorescing water. Lo and behold, the centers of the tiny green vesicles glowed red. We knew we had made bilayer membranes from nothing more than pyruvate and water.

We celebrated that night with a bottle of Napa Valley cabernet and talk of next steps and publications. We both knew that the pyruvate results were at best a footnote to the Murchison and NASA Ames discoveries, but the experiments seemed to underscore the inevitable emergence of self-organizing molecular systems along the path to life.

To be sure, many problems remain to be solved. Recent work by Deamer's group suggests that lipid self-organization may be sharply limited by the presence of dissolved calcium and magnesium, seawater ingredients that would have been present in significant concentrations in Earth's early ocean. Perhaps life can begin only in fresh water, or maybe some as yet unidentified varieties of lipid molecules were involved. And, as many biologists have been quick to point out, the vesicles produced in Deamer's work are a far cry from actual cell membranes, which feature a mind-boggling array of protein receptors that regulate the flow of molecules and chemical energy into and out of the cell.

These details will occupy researchers for decades to come, but the emergence of cell-like vesicles from simple molecules is now one of the best-understood features of life's origin.

AEROSOL LIFE

New ideas about the emergence of self-organized molecular systems keep origin-of-life workers on their toes. An especially intriguing recent proposal comes from Oxford chemist Christopher Dobson and his collaborators at the National Oceanic and Atmospheric Administration (NOAA) in Boulder, Colorado. In 2000, they published a speculative yet persuasive hypothesis on lipid self-organization in the *Proceedings of the National Academy of Sciences*. Elaborating on earlier unpublished work by the geophysicist Louis Lerman at Stanford, Dobson's group focused on the possible roles of atmospheric aerosols in prebiotic synthesis and molecular organization.

Many organic molecules—especially lipid molecules like the ones Deamer isolated from the Murchison meteorite—could have accumulated at the ocean's surface like an oil slick. As wind kicked up whitecaps and waves crashed onto the earliest shores, a continuous fine mist of aerosol particles—tiny droplets, some smaller than a thousandth of an inch across—would have sprayed into the atmosphere from the oily surface. Each water droplet would have contained a significant concentration of organic molecules that almost immediately would have formed a membranous shell around the wet interior. The largest of these droplets would have fallen quickly back into the foam, but smaller aerosol particles are quite robust and could have remained suspended in the atmosphere for months or even years, riding wind currents like microscopic gliders high into the stratosphere.

Dobson and colleagues speculate that lipids in each aerosol particle formed a spherical, single-layer structure with the hydrophobic ends facing the atmospheric exterior and the hydrophilic ends facing the aqueous interior. Many of these aerosol particles would have incorporated reactive, water-soluble organic molecules, which might have undergone further chemical reactions in sunlight. Each particle would have had weeks or months to experience such energetic transformations; each would have been, in effect, a tiny chemical experiment.

For hundreds of millions of years, aerosol particles in numbers beyond imagining drifted into the skies. Upon their return to the ocean, each hydrophobic aerosol particle would have been spontaneously coated by more lipid molecules at the ocean's surface to form a bilayer structure—the emergence of the familiar membrane structure of cellular life. In the words of Dobson, "Organic aerosols offer more than freedom from the tyranny of the tidal pool or Darwin's 'warm little pond'; they offer a possible mechanism for the precursory production and the subsequent evolution of populations of cells."

In either scenario, whether in the form of wind-blown aerosols or water-bound vesicles, lipid self-organization seems to have been an essential step in isolating the insides from the outsides of cell-like struc-

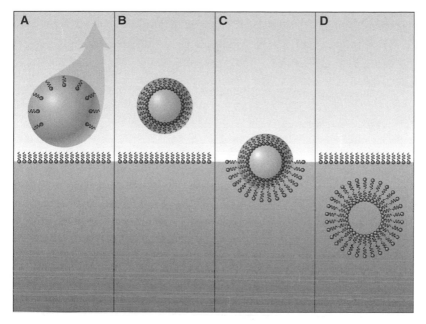

According to the theory of Christopher Dobson and colleagues, the surface of the ancient ocean was coated with amphiphilic molecules. The action of ancient waves and winds would have formed aerosol particles surrounded by lipid molecules (A). These particles might have remained in Earth's atmosphere for months (B), but would eventually return to the ocean (C), and form cell-like bilayer structures (D) (after Dobson et al., 2000).

tures. But a membrane, by itself, is not life. Other essential biomolecules, including proteins, carbohydrates, and nucleic acids, had to be assembled from the soup. The trouble is that the building blocks of these macromolecules—amino acids, sugars, bases—are all water soluble. By themselves, they can't self-assemble in water.

What to do? Call in the rocks.

12

Minerals to the Rescue

But I happen to know exactly how life arose; it's brand-new
news, at least to the average layman like yourself. Clay. Clay is
the answer. Crystal formation in fine clays provided the
template, the scaffolding, for the organic compounds and the
primitive forms of life. All life did, you see, was take over the
phenotype that crystalline clays had evolved on their own.
                              John Updike, *Roger's Version*, 1986

The first living entity emerged from interactions of air, water, and rock—the same raw materials that sustain life today. Of these three chemical ingredients, rocks—and the minerals of which they are made—have generally received little more than a footnote in theories of life's emergence. The atmosphere and oceans have long enjoyed the starring roles in origin scenarios, while rocks and minerals sneak in and out as bit players—or simply as props—and then only when all other chemical tricks fail.

Some recent and fascinating experiments promise to change that misperception. Origin-of-life researchers have begun to realize that minerals must have played a sequence of crucial roles, beginning with the synthesis of biomolecules and during their subsequent assembly into growing and evolving structures.

The Miller–Urey chemical process works by ionizing gas molecules—blasting them with lightning or ultraviolet radiation and thereby stripping off electrons, so that small groups of atoms readily recombine into larger organic molecules. Interesting molecules inevitably emerge, but those energetic processes effectively prevent the formation of essential macromolecular structures, including the polymers and membranes required by all known life-forms. The Miller–Urey scenario can't have been the entire story. That's why many researchers, especially those trained in geology, turn to rocks and minerals.

MINERALS AS PROTECTION

Rocky outcrops and overhangs—especially in tidal zones, where seawater evaporates and thus concentrates organic molecules—might have promoted macromolecular formation. Imagine a shaded cove where increasingly concentrated mixtures of organic molecules accumulated and reacted, protected by a rocky ledge from the Sun's harmful radiation. Rocks might have served as Earth's earliest sunblock. They may well have provided protection at a smaller scale as well. Many volcanic rocks of the early Earth were laced with countless air pockets left by expanding volcanic gases. Evaporating seawater might have deposited a rich mix of organic molecules in such tiny hollows, each like a small test tube where further reactions could proceed.

Mineralogist Joseph V. Smith, professor emeritus at the University of Chicago, envisions even smaller protected environments. He cites electron microscopy studies of weathered mineral surfaces, which often display myriad microscopic cracks and pores. Feldspar, the commonest of all rock-forming minerals, sometimes features millions of tiny weathered pockets, each the approximate size and shape of living cells, each providing a place for molecules to congregate, each pore and crack a separate experiment in molecular self-organization. [Plate 7]

POLYMERIZATION ON THE ROCKS

The production of macromolecules requires two concerted steps: The correct molecules must first be concentrated and then organized into the desired structure. In the case of lipid membranes, these two tasks occur virtually simultaneously and spontaneously; lipids in water separate and self-organize into a bilayer. But other key biological macromolecules, including proteins and carbohydrates, form from water-soluble units—amino acids and sugars. Consequently, they tend to break down, not form, in water.

One promising way to assemble such molecules from a dilute solution is to concentrate them on a surface. For decades, the prevailing paradigm has been that the molecules of life assembled at or near the ocean–atmosphere interface. The surface of a calm tidal pool, or perhaps a primitive slick of water-insoluble molecules might have done the job. But then, as noted, these environments are open to lightning storms and ultraviolet radiation.

Origin scientists with a penchant for geology have long recognized that rocks might provide attractive alternative surfaces for concentration and assembly—a kind of scaffolding for the assembly of protolife. More than a half-century ago, the British biophysicist John Desmond Bernal advocated the special role of clays, which are ubiquitous minerals with regularly layered atomic structures.

Clays come in a wide range of compositional and structural variants, but all of them feature layers of strongly bonded silicon and aluminum atoms. The proclivity of clays to exhibit a surface electrostatic charge enhances their ability to adsorb organic molecules—a kind of molecular-scale static cling. What's more, clays tend to occur as exceptionally fine-grained flat particles. Consequently, a palm-sized pile of ordinary clay can boast a reactive surface area greater than 1,000 square feet.

Subsequent experiments have supported Bernal's speculations. In a 1978 study, Israeli biochemist Noam Lahav and colleagues discovered that amino acids concentrate and polymerize on clays to form small, protein-like molecules. Such reactions occur when a solution containing amino acids evaporates in the presence of clays—a situation not unlike the evaporation that dries up a shallow pond or tidal pool. Of special note is the fact that this process relies on cycles of heating and evaporation—and cycles, recall, are one of the key factors in the emergence of complexity. Patterns of daily and seasonal changes doubtless fostered the emergence of new molecular structures.

More recently, research by NASA-sponsored teams in California and New York has demonstrated that a variety of layered minerals can adsorb and assemble a variety of other organic molecules. In a *tour de force* series of experiments during the past two decades, chemist James Ferris and colleagues at Rensselaer Polytechnic Institute induced clays to act as scaffolds in the formation of RNA, the polymer that carries the genetic message enabling protein synthesis.

Ferris relied on the simplest of procedures. First, he prepared a solution of "activated" RNA nucleotides, each consisting of a ribose sugar bonded to a phosphate and a base, plus a reactive molecule called imidazole that promotes, or "activates," bonding between nucleotides. Such a solution can sit on the lab bench for weeks with little change.

But sprinkle in a bit of a suitable clay mineral and the RNA pieces start to link up. In a matter of hours, lengths of 10 nucleotides form. By the end of 2-week experiments, the RPI team produced RNA strands of more than 50 nucleotides. The fine-grained clay particles had induced polymerization by a process not yet fully understood.

Buoyed by the Ferris team's success, other origin-of-life researchers tried their hand at other biopolymers. Leslie Orgel, research professor at the Salk Institute for Biological Studies in San Diego, succeeded in forming a variety of proteinlike chains of amino acids up to several dozen molecules long. Orgel and his students discovered that different minerals preferentially select and polymerize different molecules from a water-based solution. By combining the right mineral with the right molecule, they could form polymers at will.

In conjunction with his experiments, Orgel also developed an elegant theory of "polymerization on the rocks," in which he pointed out both the promise and problems with mineral surfaces. Minerals such as clays and hydroxides certainly can adsorb interesting biomolecules, he noted, including the amino acids and nucleotides essential to life. Furthermore, once two of these molecules are adsorbed close to each other, they have a tendency to bond. As more and more molecules are added to a lengthening polymer, however, the strand becomes more and more tightly bound to the mineral surface. How, he asks, can a polymer contribute to life if it's stuck to the rocks?

One possible answer came from the Harvard University laboratory of geneticist Jack Szostak, who mixed together clays, RNA nucleotides, and lipids in the same experiment. Lo and behold, the clays not only adsorbed RNA, but also hastened the formation of lipid vesicles. In the process, RNA-decorated clay particles were incorporated into the vesicles. This spontaneous self-assembly of RNA-containing vesicles, though a long, long way from synthesizing life, is perhaps the closest anyone has come to forming a cell-like entity from scratch.

MORE MINERAL MAGIC

Every interaction between a mineral and a molecule requires knowledge of two different chemical entities—a crystalline solid and a tiny carbon-based cluster of atoms. Most origin-of-life experts come from the world of organic chemistry—the carbon-based biomolecule part. It's easy to get the impression that minerals are brought into origin-of-life experiments only when nothing else works, and then as a sort of magic powder. Atomic-scale details of the molecule–mineral interactions are usually fuzzy at best.

Gustaf Arrhenius, a senior NASA-supported researcher at the Scripps Institution of Oceanography, in La Jolla, California, has a somewhat different background. He knows organic chemistry, to be sure, but he was trained in inorganic chemistry and has a deep appreciation of crystalline solids and their structural idiosyncrasies. Consequently, he has zeroed in on the mineral group known as double-layer hydroxides. Like clays, the double-layer hydroxides are common in nature, and they come in almost limitless compositional variants, with magnesium, iron, chromium, nickel, calcium, aluminum, and many other elements mixing and matching in the double-layer structure. Additional complexity arises because small molecules—water, carbon dioxide, or a variety of other common species—occupy the spaces between the layers. By changing their chemical contents, Arrhenius has been able to fine-tune his double-layer hydroxides to perform specific chemical tasks.

Unlike Ferris and Orgel, Arrhenius did not try to form long polymers on the surfaces of his mineral powders; rather, he exploited the tendency of double-layer hydroxides to soak up small organic mol-

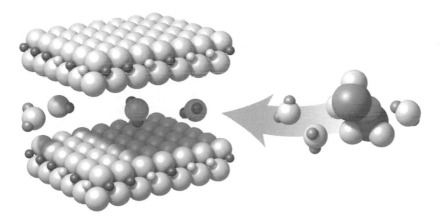

Gustaf Arrhenius demonstrated that double-layer hydroxide minerals have the ability to attract small molecules and catalyze their reactions into larger species of biological interest (after Arrhenius et al., 1993).

ecules in the rather large spaces between the layers. Once confined and concentrated, these small molecules tend to form larger molecular species that are not otherwise likely to emerge from the soup. His most interesting products so far have been sugar phosphates, in which a sugar molecule is linked to a phosphate group—a suggestive pairing, in that it forms the backbone of every DNA and RNA molecule.

Joseph Smith has explored yet another intriguing mineralogical wrinkle. Smith is an authority on zeolites, a diverse family of natural and synthetic crystals that feature latticelike frameworks of silicon, aluminum, and oxygen atoms. These open frameworks result in molecule-sized channels that run the length and width of the crystals. The tendency of some hydrocarbon molecules to enter these zeolite channels, while other molecules remain behind, lies at the heart of the multitrillion-dollar petroleum refining business.

Instead of looking at mineral surfaces, Smith imagines prebiotic molecules concentrating inside zeolite channels. Zeolites abound in the weathering products of volcanic rocks; organic molecules were certainly present in the same environments. Once packed with a long column of molecules, a zeolite might promote additional reactions, including polymerization. Smith even suggests that the first cell wall might have been "an internal mineral surface." Experiments have yet to back up these ideas; it's hard to study what's going on inside a mineral. But the message is clear—minerals and molecules could have interacted in a lot of intriguing ways on the early Earth.

CLAY LIFE

Many researchers have looked to minerals to give life a jump-start, whether as protective containers, organizing surfaces, or catalytic engines of prebiotic chemistry. In a provocative 1966 article (and in numerous subsequent publications) the Scottish origin-of-life expert A. G. (Graham) Cairns-Smith carried these ideas even further by speculating that fine-grained crystals of clay might themselves have been the first life on Earth.

Virtually all other authorities assume that life emerged by a continuous chemical process of increasingly complex carbon chemistry. Today's biochemistry is thus a direct, albeit highly evolved, successor of ancient carbon geochemistry. Cairns-Smith disagrees. "I believe the unity of biochemistry is of no direct help," he says. "Evolution did not start with the organic molecules that have now become universal to life: indeed I doubt whether the first organisms . . . had any organic molecules in them at all." He points to five decades of failed attempts to synthesize key macromolecules, such as RNA and proteins, in any plausible prebiotic scenario. In his view, nucleic acids and proteins are too complicated to build without help from the inorganic world.

Cairns-Smith's ideas at first seem fanciful, but he is a persuasive scientist, writer, and lecturer. He speaks in a soft Scottish accent, engaging his audience with a lucid, low-key delivery. He's not pushy with his ideas, but he has the knack of drawing you into his way of thinking. His writings, both for academic and general audiences, eschew jargon and rely on simple, vivid illustrations. *Seven Clues to the Origin of Life*, his popular book on the clay-life hypothesis, reads like a mystery novel, complete with dialog, false leads, and quotations from Sherlock Holmes. Whether you're persuaded or not, you have to admire Cairns-Smith's style.

The crux of the argument rests on a simple analogy. Cairns-Smith likens the origin of life to the construction of a stone archway, with its carefully fitted blocks and crucial central keystone that locks the whole structure in place. But an arch cannot be built simply by piling one stone atop another. "The answer," he says, "is with a scaffolding of some kind." A simple support structure facilitates the construction and can then be removed. "I think this must have been the way our amazingly 'arched' biochemistry was built in the first place," he wrote in a *Scientific American* article in 1985. "The parts that now lean together surely used to lean on something else—something low tech." That something, he suggests, was a clay mineral.

Like Bernal before him, Cairns-Smith points to the ubiquity of clays, their negative surface charge, their affinity for organic molecules, and the immense reactive surface area their fine-grained layers provide. As he was quick to point out in a Geophysical Lab seminar in the summer of 2003, "I'm an organic chemist, not a clay mineralogist—though I often talk like one." Then he embellishes the point with a surprising twist that reveals his mineralogical sophistication. Clay crystals grow, and their mutable layered structure can carry and pass on a kind of "genetic information," in the form of a reproducible sequence of crystal defects or variable chemical compositions. Given their ability to grow and reproduce, might not clays be thought of as living?

Even the most chauvinistic of mineralogists tend to balk at Cairns-

Smith's concept of "clay life" and "crystal genes," but he does have an intriguing point. Genetic information in modern life is carried by DNA and RNA molecules, which, as noted, are each made up of sequences of only four different nucleotides—in effect, a four-letter genetic alphabet. Any conceivable message can be conveyed with such an alphabet and an appropriate code.

In much the same way, clay minerals commonly display periodic structural or compositional variations that might be used to convey information. For example, the clay structure features strong sheetlike layers of atoms that stack one on top of another. Each of these structural layers can adopt one of three different orientations relative to its neighbors—at 0, 120, or 240 degrees. The sequence of layer orientations is analogous to a three-letter alphabet. What's more, layers of different thickness and chemical character sometimes interleave with each other, adding dramatically to the information-carrying potential of clays.

Periodic variations can also occur *within* any given layer; that is, each layer can display a feature called "twinning," by which a given layer can incorporate small regions in all three different orientations. What's more, many clays boast complex chemical compositions, with varying mixtures of aluminum, magnesium, iron, and other elements. These atoms reside in predictable repeating sites, each one surrounded by an octahedron of six oxygen atoms. Move along the surface of the clay particle and you'll find an octahedral site every ten-billionths of an inch or so, as regular as clockwork. But these regimented sites in the clay structure often contain a more-or-less random sequence of aluminum, magnesium, and other elements, as well as vacant sites and other defects. Cairns-Smith argues that such a random distribution of atoms at the exposed mineral surface can also act as a kind of genetic sequence, each element representing a different letter in this peculiar alphabet. "In two-dimensional crystal genes, information would be held as a pattern on one face of the crystal," he posits.

So what? How can a random sequence of layer orientations or a pattern of elements at the clay surface be regarded as a living entity? How can it possibly reproduce itself? Cairns-Smith proposes two intriguing, though unproved, possibilities. First, he speculates, clay crystals might grow in such a way as to copy any given layer, stacking sequence, or element pattern over and over again, and then flake apart in what amounts to an act of self-replication. In this scenario, clay par-

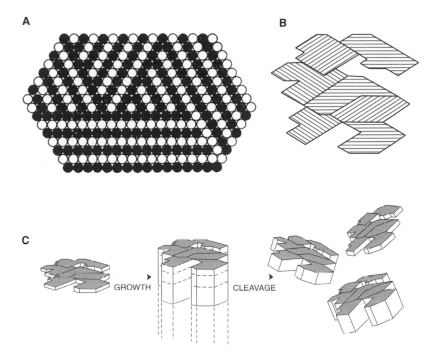

Graham Cairns-Smith suggests that clay minerals, which have a sheetlike layered structure, can carry genetic information in the form of three different layer orientations (A). Individual clay particles can possess a complex pattern of these orientations (B). If sequences of layers can be copied, then this genetic information might be passed on from one generation of clays to the next (C) (after Cairns-Smith, 1988).

ticles crystallize over and over again, and particularly stable sequences of layers or elements ultimately win out in a mineralogical version of evolution by natural selection.

Moreover, according to Cairns-Smith's model, different exposed layer edges or elements at the clay surface have affinities for specific organic molecules. Not only do the negatively charged clay surfaces readily attract organic molecules, but clays also have the ability to catalyze reactions between surface-bound molecules. The self-replicating inorganic world of clays could have acted as the long-abandoned scaffolding for the organic world—a process he calls "genetic takeover." The resulting macromolecule might then possess exactly the sort of information-rich structure needed for the emergence of a carbon-based genetic mechanism.

TESTING

Any acceptable scientific theory must make testable predictions; otherwise, as Karl Popper consistently maintained, a theory is just idle speculation. Graham Cairns-Smith is well aware of this requirement, and he does not shy away from predictions.

A potentially testable feature of the clay-life hypothesis follows from the central requirement that clays evolve. In 1988, Cairns-Smith wrote a provocative essay entitled "The Chemistry of Materials for Artificial Darwinian Systems," in which he elaborated on his idea that clays carry "genetic information" in the form of variable crystal patterns that can be passed from one generation of crystals to the next. Clays continuously form and dissolve in many geological environments. Perhaps, he posited, clays evolve by the selective replication of favorable patterns, coupled with the selective dissolution of unfavorable patterns. In this way, the evolution of clays is analogous to that of microorganisms. "The first step is to grow up a batch of the organisms," Cairns-Smith explains. "Then a small sample of this is used to seed a new batch, which is in turn grown up—and so on. In such circumstances those types which reproduce fastest will be most likely to be carried on between the batches and will eventually become the only types present."

By analogy, Cairns-Smith predicts, more successful (that is, more likely to be replicated) element arrangements will gradually dominate less successful variants. But can we test this idea in a laboratory experiment? Like a study of microorganism evolution, we would have to characterize the detailed state of numerous particles in the first generation of clays, and then monitor changes in subsequent generations. Cairns-Smith recognizes many experimental challenges. "Can the material be synthesized in the laboratory on a practical time scale? Can we find conditions that will allow crystals to grow? . . . And then the $64,000 one: Do sequences replicate accurately enough through growth?" Underlying all of these questions is the technical barrier: how to determine the exact elemental sequence of a clay particle.

For microbial populations, these problems have been solved. Microbes grow rapidly, they divide predictably, and they copy their DNA with fidelity. And it's relatively easy to conduct a broad survey of microbial DNA sequences—a task that has become automated by decades of genetic research and billions of dollars in capital investment. Molecular biologists have thus learned to document evolution in a microbial community.

But for clays, no such sequencing technology exists. To be sure, modern imaging instruments can provide some insights. Cairns-Smith advocates high-resolution transmission electron microscopy (HRTEM): "We should go for materials for which we can expect HRTEM to be applicable," he writes. Nevertheless, characterization of the element arrangement in a single clay particle, much less the thousands of separate analyses necessary for an adequate clay-particle population survey, is beyond any current technique. This situation is nothing new—scientific progress has often had to wait for new technology.

Many earth scientists are drawn to the concept that clays, or some other mineral, played a crucial role in biogenesis. There's a satisfying completeness in the integration of the solid, liquid, and gaseous parts of our planet to make life. Nevertheless, I find the reliance on clays a bit unsettling. It seems that all a chemist needs to do is sprinkle in a little clay powder and *voilà*: polymerization, self-assembly, synthesis, stabilization! Clays seem to do it all. The phenomenology is fascinating, but what's actually going on at the atomic scale? Clays are inherently variable in their composition, ill-defined in their surface properties, and so fine-grained that it's virtually impossible to be sure what molecule is binding to what surface. The "magic powder" aspects of clays leave me frustrated.

Given the choice between a poorly characterized fine-grained powder and an elegant faceted single crystal, I'll take the crystal any day. And, as it turns out, that's the *only* way you can tell left from right.

13

Left and Right

*Assemblage on corresponding crystal faces of enantiomorphic
pairs of crystals, such as right-hand and left-hand quartz
crystals, would provide us with a most simple and direct
possibility of localized separate assemblage of right- and
left-hand asymmetric molecules.*

Victor M. Goldschmidt, 1952

Prebiotic processes produced a bewildering diversity of molecules.
Some of those organic molecules were poised to serve as the essential starting materials of life—amino acids, sugars, lipids, and more.
But most of that molecular jumble played no role whatsoever in the
dawn of life. The emergence of concentrated suites of just the right
mix remains a central puzzle in origin-of-life research.

One of the stages of life's emergence—an early and confounding
one at that—must have been the incorporation of handedness. Many
of the most important biomolecules, amino acids and sugars included,
come in mirror-image pairs: something like your two hands, which
have the exact same structure, but can't be exactly superimposed. In
the same way, pairs of these so-called chiral molecules have the exact
same composition and structure, and many of the same physical properties, but they are mirror images of each other.

Virtually all known prebiotic synthesis pathways produce chiral
molecules in 50:50 mixtures. No obvious inherent reason exists why
left or right should be preferred. And yet living cells display the most
exquisite selectivity, choosing right-handed sugars over left and left-
handed amino acids over right. In spite of a century and a half of study,
the origin of this biochemical "homochirality" remains a central mystery of life's emergence.

THE HANDEDNESS OF LIFE

Molecular chirality arises from a common circumstance of organic molecules. In many molecules, one carbon atom forms four bonds to four *different* groups of atoms. The amino acid alanine, for example, has a central carbon atom linked to one NH_2 molecule (an amino group), one COOH molecule (a carboxyl group), one CH_3 molecule (a methyl group), and one lone hydrogen atom. If you orient the hydrogen atom up and look down on the molecule, there are two ways to arrange the remaining three components. So-called "right-handed" or D-alanine has a clockwise arrangement of carboxyl–amine–methyl groups. Chemists employ the letter "D" after the Greek "dextro" for "right" (though the designation of molecular "right" vs. "left" is a completely arbitrary convention.) Alternatively, "left-handed" or L-alanine features a clockwise arrangement of amine–carboxyl–methyl groups. In this case the "L" comes from the Greek "levo," for "left."

For reasons that are still not well understood, life selects L-amino acids and D-sugars almost exclusively over D-amino acids and L-sugars. One critical consequence of this selection is that our cells often respond very differently to other kinds of chiral molecules. The familiar flavoring limonene, for example, tastes like lemons in its right-handed form, but like oranges in the left-handed variant. More sinister is the behavior of the infamous drug thalidomide: The left-handed form cures morning sickness, while the right-handed form induces birth defects. Consequently, the Food and Drug Administration demands

Many biomolecules occur in both left-handed and right-handed variants, which are mirror images of each other. This situation arises when a central carbon atom is linked to four different groups of atoms—groups that can be arranged clockwise or counterclockwise.

chiral purity in many pharmaceuticals—a difficult processing step that adds more than $100 billion annually to our drug costs.

Chirality is an essential, diagnostic characteristic of cellular life. But how did this selectivity emerge in the seemingly random prebiotic world? A dozen theories, from the mundane to the exotic, have been proposed, but all ideas fall into one of two general categories.

Some experts suspect that prebiotic synthesis was an inherently asymmetrical process, leading to an inevitable global excess of L- over D-amino acids on the prebiotic Earth. More than a century and a half ago, Louis Pasteur demonstrated that left- and right-handed crystals cause polarized light to be rotated in opposite directions. Conversely, the orientation of polarized light shining on a chemical-rich environment might influence the relative proportions of left- versus right-handed molecules, either by selective synthesis of L-amino acids, or by selective breakdown of D-amino acids. Though the effect is generally small, this kind of chiral-selective process might conceivably have occurred in deep space near a rapidly rotating neutron star, or perhaps on Earth's surface as the result of polarized sunlight reflected off the ocean surface. No one knows for sure what hundreds of millions of years of such processing might yield.

Other physicists echo this theme, but posit that asymmetric synthesis resulted from so-called parity violations that occur during radioactive beta decay. In physics, the parity principle states that physical processes appear exactly the same when viewed in a mirror. Beta decay, which occurs when an unstable radioactive atom loses an electron (or a positron) in the process of becoming a stable atom, is the only known physical event that violates this parity principle. Beta-decay events produce polarized radiation of only one handedness, and this chiral radiation, in turn, could have enhanced the synthesis of L- versus D-amino acids. One problem with these asymmetric processes is that they seem to yield only the slightest excess of one handedness over the other—generally less than a minuscule fraction of 1 percent. Such minute effects hardly seem sufficient to tip the global balance toward L-amino acids or D-sugars.

LOCAL SYMMETRY BREAKING

Many origin-of-life researchers (myself included) argue that chiral selectivity more likely occurred as the result of an asymmetric local, as opposed to global, physical environment on Earth. After all, the emergence of life consisted of a series of chemical events, each of which occurred at a specific place and a specific time. It is very possible that those emergent steps were repeated countless millions of times across the globe, but each individual emergent event was local: It occurred at specific location with specific molecules. Consequently, if the local environment where a reaction occurred was strongly chiral, then chiral molecules might have emerged.

The chemical process of life's origin is in some respects like the formation of a crystal. In life, as in crystals, two essential and largely independent steps are necessary. The first step in crystal formation, nucleation, requires the precise organization of a relatively small number of atoms or molecules into a "seed." This seed might by chance have either a D or L character. Then comes growth, as the original seed provides a template for the ordered assembly of more atoms and molecules. Each step in the chemical origin of life must also have required nucleation, followed by growth.

In life, as in crystal formation, these two stages usually proceed at very different rates. Nucleation may take place with ease, while growth is slow. Such a situation leads to the myriad microscopic crystallites that give colorful agates and frosted glass their distinctive translucent optical properties. If, on the other hand, nucleation is rare while growth is rapid, then a single large crystal may form.

Since life is vastly more complex than any crystal, it is reasonable to think that nucleation—the self-organization of molecules into a replicating entity—is relatively rare, perhaps even a singular event in Earth's history (though that is by no means a certainty). But, once formed, this protolife must have grown rapidly, in the process consuming every available molecular feedstock and frustrating further origin events. If that life-form was by chance homochiral, D or L, then that handedness would be passed on to subsequent generations. All it takes is an initial local chiral environment.

We now realize that such local environments abounded on the prebiotic Earth. Indeed, every chiral molecule is itself a tiny local chiral environment that might select other molecules of similar handedness. The pioneering chirality studies of Louis Pasteur relied on this characteristic: When he evaporated a 50:50 solution of D- and L-tartaric acid, the mixture spontaneously divided into pure D and pure L crystals. Such a circumstance arises because D-D or L-L pairs of tartaric acid molecules happen to fit together more easily than D-L pairs.

Might prebiotic organic molecules have displayed the same behavior? Evidence is still spotty, but a few experiments do reveal a strong tendency for chiral self-selection in certain polymers. The Swiss chemist Albert Eschenmoser, who explored the stabilities of more than a dozen variants of RNA with modified sugar-phosphate backbones, has found that some (but not all) of these molecular chains grow spontaneously with greater than 90 percent chiral purity. Perhaps prebiotic polymers became homochiral by a similar self-selection process.

One caveat: In a dilute, complex primordial broth, the chances of linking together *any* two types of molecule, much less two useful monomers of the same handedness, would have been exceedingly small. That's why James Ferris, Leslie Orgel, Graham Cairns-Smith, and others have resorted to mineral surfaces, which induce polymerization by concentrating and aligning desirable molecules. But some mineral surfaces have an added advantage. Every rock, every grain of sand or particle of silt also has the potential to offer chiral environments, in the form of asymmetric mineral surfaces. Left- and right-handed mineral surfaces might provide the perfect solution for concentrating and separating a 50:50 mixture of L and D molecules.

Minerals often display beautiful crystal faces, which might have provided ideal templates for the assembly of life's molecules. A few minerals, most notably quartz (the commonest grains of beach sand), occur in both right-handed and left-handed structural variants. The quartz structure features helices of atoms that in some crystals spiral to the left and in other crystals to the right. Every grain of beach sand thus provides a chiral environment.

But even though the vast majority of minerals are "centrosymmetric," and thus not inherently handed, their crystals commonly feature pairs of faces whose surface structures are mirror images of each other. Like quartz, these chiral surfaces have arrangements of atoms that are ideally suited to select and concentrate L versus D molecules, such as amino acids or sugars.

Natural left- and right-handed surfaces occur in roughly equal numbers, so there's not much chance of chiral selection occurring on a global scale. But, once again, here's the key: The chemical origin of life was not a global event. The first common ancestor—the precursor to

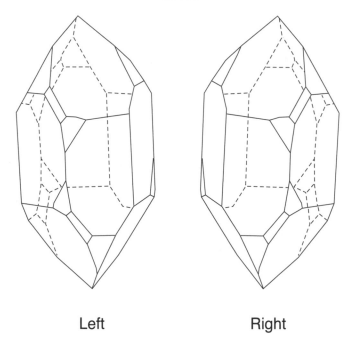

Left   Right

The common mineral quartz forms both left-handed and right-handed crystals.

all of the varied life-forms we see on Earth today—arose as a bundle of self-replicating chemicals at a specific place and time. Once that chiral system began making copies of itself, the handedness of life was fixed. To many experts in the field, the choice of L- versus D-amino acids seems to have been a one-time, chance event, and that scenario points to a simple mineralogical mechanism for chiral selection.

CHIRAL MINERAL SURFACES

As a mineralogist for more than a quarter of a century, I love the idea that crystals may have played a central role in life's beginnings. For a time, the question of origins took me away from minerals, but they have called me back to some of the most delightful experiments of my career.

The chirality problem represents a particularly puzzling aspect of the more general question of molecular selection, but it's also an aspect that can be studied with well-controlled experiments. We know that prebiotic synthesis processes yield huge numbers of different molecules, of which life uses relatively few. It would be impossible to study that full range of prebiotic products in one experiment. But using pairs of chiral molecules makes things a lot easier: Because all known prebiotic syntheses result in equal amounts of left- and right-handed molecules, it's relatively easy to design an experiment that starts with a 50:50 mix and looks for environments that separate left from right. If we can discover the processes by which handed molecules were selected and concentrated, then there's hope that we can solve the more general selection problem as well.

For most of the twentieth century, scientists have recognized the power of minerals to select molecules—processes in which one kind of molecule sticks more strongly to a surface than another. If two molecules differ significantly in size and shape, then such selection is easy to comprehend; it's just a matter of which molecule fits best. But selection between two mirror-image molecules is more difficult. Such pairs of molecules are chemically identical, so each type of molecule forms exactly the same kinds of bonds with the mineral surface. The only way for chiral selection to occur is for the molecule to form *three* separate points of attachment, and those three bonds can't be in a straight line. You've experienced this kind of selection process if you've ever gone bowling. The ball has three holes, for your thumb and second and third fingers. A left-handed bowler can't use a right-handed ball and vice versa, because the three holes aren't in a line.

By the 1930s, scientists in Greece and in Japan had applied these ideas and tried separating left- and right-handed molecules by pouring D- and L-solutions over left- or right-handed quartz. During the next half-century, a dozen similar experiments were attempted. The basic idea was sound: Different-handed molecules do have the potential to stick selectively to different-handed surfaces. Nevertheless, these early experiments were flawed. In an effort to maximize the surface area of interaction, and thus the magnitude of the desired effect, the scientists ground their beautiful quartz crystals into fine powder. Powdering increases the surface area of a peanut-sized crystal a thousand times or more, but it destroys the flat crystal surfaces that might promote selection. A powder displays every possible crystal surface simultaneously. Some of these surfaces may very well select L-amino acids with great efficiency, but surfaces with a different exposed atomic structure might just as likely select D-amino acids. Even if an experiment revealed a small preference for L or D molecules, there would be no hope at all of determining which crystal surface was doing the selecting.

I decided to try a different approach.

Big crystals are the key—fist-sized crystals with fine flat faces. That's the only way to understand the atomic-scale interaction between molecule and crystal. But what mineral fits the bill?

My crystals had to be big because a layer of molecules adsorbed onto a surface an inch square weighs at most a few billionths of a gram—a daunting analytical challenge. To measure that effect, I had to find crystals with faces at least a few square inches in area. Big crystals of most minerals tend to be astronomically expensive, thanks to the voracious appetite of mineral collectors; so I had to find a common rock-forming mineral that collectors don't covet. As an added benefit, the commonest minerals are also likely to be the most relevant to origin scenarios. Most important, the crystals had to possess faces with surface structures that lack mirror symmetry. Only a chiral face could accomplish the chiral selection task.

I thumbed through my favorite mineralogy book, a frayed, dog-eared copy of Edward Dana's *A Textbook of Mineralogy*, purchased at the American Museum of Natural History when I was 14. Classic line drawings of crystal forms decorate almost every page. The vast majority of crystal faces, I realized, aren't chiral. Many of the commonest minerals—garnet, olivine, mica, pyrite—won't do.

Then I turned to page 512—calcite, an abundant mineral and the one most closely associated with life. Calcite is the mineral of clamshells and snail shells, of pearls and coral. Lo and behold, its commonest crystal faces are chiral. Most calcite crystals feature a graceful, six-sided, pointed form with the fancy scientific name of scalenohedron. Everybody in the business calls it a dogtooth.

A quick search of eBay confirmed that dogtooth calcite crystals are both common and cheap. I bid on three pretty specimens from the then thriving (but now defunct) Elmwood lead-zinc mine outside Carthage, Tennessee. A week and $40 later I had the beginnings of what

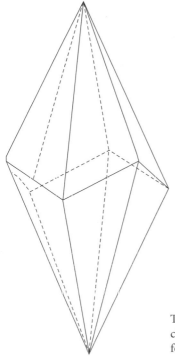

The commonest crystal shape of the mineral calcite is the so-called "dogtooth." All 12 faces of this
form are chiral.

has become a sizable calcite crystal collection. Now to design an experiment.

I wanted to find out whether chiral crystal surfaces would selectively
attract chiral molecules. I had fist-sized calcite crystals with left- and
right-handed faces, and I had D- and L-amino acids. Would D- or L-
amino acids attach themselves preferentially to left- or right-handed
calcite faces?

   Seventy years of false claims and ambiguous results had set the bar
high; a casual, sloppy experiment wouldn't do. I knew that many previous authors had started with concentrated D- or L-molecules, and then
looked for a difference in behavior as those molecules interacted with a
chiral crystal (usually quartz). That experimental path was fraught with

difficulties, since every nook and cranny of the environment is already contaminated with life's excess L-amino acids and excess D-sugars. If you look for a chiral effect, you're probably going to find a chiral effect.

Instead, I prepared a 50:50 solution of D- and L-aspartic acid, an amino acid known to have special affinity for calcite in the proteins that add strength and resiliency to all sorts of shells. The idea was to expose both left- and right-handed calcite faces to this mixed solution, and then see which molecules stuck to which face. If crystal faces really do have the ability to select chiral molecules, then left- and right-handed calcite faces should attract equal and opposite excesses of L- and D-molecules.

The experiment sounded easy in principle, but it took months to get right. First we had to figure out a way to clean the crystals, which had been handled by dozens of people, from Tennessee miners to Geophysical Lab scientists. The slightest contamination—a fingerprint, a breath, even a speck of dust—could skew our results. We settled on a procedure of repeatedly washing each crystal in a sequence of strong solvents.

In the first failed round of experiments, postdoc Tim Filley (now a professor at Purdue) and I glued little plastic reservoirs directly onto the calcite crystal faces in hopes of minimizing contamination from dust, bacteria, and other chiral influences in the air. After some practice, we got the little plastic pieces to stick to the smooth calcite surfaces. What we hadn't realized is that our epoxy glue and our plastic containers were absolutely loaded with chiral molecules, which are part of the plastic-making process. Our careful procedure introduced far more chiral contamination than the atmosphere ever could.

Simple experiments are usually best, so I resorted to what seemed like an almost childish approach. I poured the 50:50 aspartic acid solution into glass beakers and just dunked the lustrous fist-sized calcite crystals into the liquid. Twenty-four hours seemed like a good amount of time to let the crystals soak. I came back the next day, removed the crystals, washed them off quickly in pure water, and then ever so carefully extracted the attached molecules from the crystal surfaces. It was an easy task, because calcite dissolves readily in dilute hydrochloric acid. I held a crystal in my gloved left hand and oriented one face parallel to the bench top. With my right hand, I used a long dropper to cover that face with a thin puddle of acid. We were fortunate that water beads strongly on calcite, so it was relatively easy to keep the acid from spilling over the sides. After about 20 seconds, I sucked the acid back up into the dropper and emptied it into a waiting vial. We knew that any attached molecules would have been stripped off along with the outer layers of calcite.

Face by face, I applied the acid wash to each of four crystals—more than 20 surfaces in all. Most of these faces were left- or right-handed, but a few were flat, fracture surfaces—the natural breakage planes of the crystal—that are not chiral at all. Those faces would serve as a control on our methods, for they should show no preference for D- or L-amino acids. After an hour and a half of concentrated effort, we had 23 vials in hand. Would they show the desired effect?

In spite of the hassles with contamination, generating the amino acid samples turned out to be the easiest experimental task by far. The real trick, mastered by only a handful of analytical chemists in the world, was to determine whether the calcite crystal faces had preferentially selected D- or L-aspartic acid.

Tim Filley is an expert at organic chemical analysis, so we were confident that we could do the work ourselves. We had an adequate gas chromatograph/mass spectrometer, the analytical machine of choice; we had a chiral chromatographic column that should separate D- from L-aspartic acid; and we had a well-documented procedure for preparing our samples prior to injecting them into the GCMS. The setup took some time, but the methods seemed straightforward.

After a lot of practice with aspartic acid standards in different D:L ratios and different concentrations, we were ready to try the real thing. Our first analyses took place just before Christmas 1999.

GCMS affords a kind of rapid (if not instantaneous) gratification, because the ratio of two compounds, such as D- and L-aspartic acid, corresponds to the areas of two peaks that appear side by side on a graph. Joined by Tim's wife, Rose, who is also a skilled analytical chemist, we injected our first sample and waited. It took the better part of 20 minutes for the molecules to work their way through the 50-meter-long coiled chiral column, which was packed with chemicals that slow down L-amino acids slightly more than D. D-aspartic acid passed through the column about 20 seconds faster than L-aspartic acid. The first peak showed up sharp and clear—plenty of D-aspartic acid molecules there. The next 20 seconds seemed to take a lot longer than any 20 normal seconds should, but the second peak, representing L-aspartic acid, appeared on schedule. And it was clearly bigger—at a guess, more than 10 percent bigger. We allowed the machine to run just a few minutes longer, to make sure there weren't any obvious contaminants, and then stopped to let the computer calculate the peak areas.

The actual ratio turned out to be closer to a 5-percent excess of L, but it was a tantalizing result that clearly pointed to chiral selection. We decided to run the same sample a second time, just to be sure. Twenty minutes later, the pair of peaks appeared again, but this time we saw no more than a 1-percent excess of L, not enough to be sure the effect was real. The average value was a 3-percent excess, while our experimental error was evidently close to ±2 percent. Tim suggested that we run a couple of more standards to be sure the machine was giving us clean peaks.

We were off and running, with 22 more samples, each to be prepared and run in duplicate, plus standards after every few sample injections. Given the prep time, it took more than two hours to test each sample or standard. We started running in three overlapping shifts. With two small girls, Rose was always up early and did the first few injections. Tim and I usually couldn't stay away, so we'd take over in the morning and keep things running through the evening and into the wee hours. For the first couple of days, we were there 18 or more hours at a stretch, running sample after sample, eagerly watching the screen as the pair of peaks appeared 20 minutes into each run. It was addictive and exhausting, like a slow-motion video game.

So it went: Run a few samples with tantalizing results, then more standards. One encouraging sample suggested a 4-percent excess of D-aspartic acid—a critical result, since L contamination, the life sign, was everywhere, but an excess of D could come about only through crystal selection. Still, the results weren't consistent. What was even worse, the peak shapes began to broaden and degrade after only about 25 injections. Broad peaks overlap, making analysis of the D:L ratio all but impossible.

We called the supplier of the chiral column to complain—a new $500 column like ours should have lasted for hundreds of runs, not just 25. They suggested that we bake out the column to make sure that large, unwanted molecules hadn't contaminated it. We had been turning off the machine a few minutes after we saw the aspartic acid peaks, because we were eager to do the analyses as fast as possible. Perhaps other slower molecules hadn't made it all the way through the column and were gumming up the system.

After the prescribed treatment, the column worked better for a few injections, but soon the peak shapes were worse than ever. We called the company again.

"What kind of samples are you running?" they asked.

"Just aspartic acid," Tim said.

"Are you sure there's nothing else in the samples?" they responded.

"Well, maybe a little bit of dissolved calcite; it's just calcium carbonate."

That was it. The company rep told us that the mineral residue had quickly degraded our column, making it useless for further work. They generously offered to send us another for free (probably suspecting that we'd be going through a great many $500 columns in the coming months).

So now we'd have to execute an additional chemical step to remove the calcium carbonate with acid before injection. We weren't happy about that, because each additional treatment of the samples increased the chance of contamination.

We kept trying, but those first experiments were just too flaky. Lots of L excesses of a few percent, one or two D excesses, but nothing really systematic. What's more, some of the L excesses came from the fracture surfaces, which should have had no effect, while the right- and left-handed calcite faces gave inconsistent results at best.

We had spent months at the project with nothing to show for it but a lot of experience under our belts. In the spring of 2000, Tim's postdoc was nearly up. He and Rose were looking forward to setting up a new home in Indiana. I had to make a decision whether, and if so how, to continue.

I decided to give it another shot. One thing was absolutely clear: The experiments had to be performed in ultraclean facilities, with every possible precaution against contamination. My colleagues at our sister lab, the Department of Terrestrial Magnetism (DTM for short), maintain a chemical clean lab at our Northwest Washington campus for isotope and trace-element work, and they generously let me use one of their high-tech chemical hoods for a few months.

In May of 2000, I hauled box upon box of chemicals, glassware, and crystals to the second-floor clean lab of DTM's venerable experiment building. The sparkling, well-lit labs could be entered only through a small glassed-in vestibule, where the transition to clean-room protocols took place. I removed my shoes and slipped into a pair of pale-blue booties, white coveralls, hood, respirator mask, and latex gloves. The protective gear wasn't uncomfortable, but it took a bit of getting used to.

After organizing my supplies, I began the experiment. Leaving nothing to chance, I treated the thrice-washed crystals with extra care, used the highest-purity amino acids, ultra-pure water (at $40 per gallon!), and freshly baked glassware. Every step was carried out in a chemical hood, a glass-enclosed volume about 4 feet wide, 2 feet deep, and 3 feet high—plenty of room for my beakers. My hood was maintained with positive pressure to prevent outside air from entering the enclosed area.

The experiments were conducted almost exactly as before. First, soak four crystals for 24 hours in the 50:50 amino acid bath. This time I made sure that the pH of the solution remained close to 8, a value ensuring that the calcite would not start to dissolve. Then, after a day of soaking, I repeated the now-familiar wash process, applying acid to each crystal, face by face. The day's work produced 23 small glass vials of acid extract. Over the next two days, I repeated the entire experiment two more times to be absolutely certain.

A week's work yielded 69 vials of aspartic acid solution washed from crystal faces, plus several vials with individual samples of each day's aspartic acid solution, of the pure water, and of the acid wash. I wanted samples of everything, in case I had to track down contaminants. More than 75 sample vials needed to be processed, each to be analyzed at least 3 times. Two-hundred-plus amino acid analyses is a huge job, and the Geophysical Lab facilities were not up to it. So for that crucial, final step I headed downtown, hat in hand, to George Washington University, to see geochemist Glenn Goodfriend, a former colleague at the Geophysical Laboratory.

Success in a scientific career can be measured in many ways. Some scientists crave admiration and respect from their peers. Others prize a flexible job that gives them the freedom to pursue any line of research. And for many researchers, the quiet exuberance of doing good science is the prime measure of a happy and successful career. By all these measures, Goodfriend was one of the most successful scientists I'd ever known.

At the agreed-upon late-morning hour, I arrived at his modest office in the basement of Lisner Hall, home of the Geology Department. Glenn was a research professor, his work sustained almost entirely by a succession of two- and three-year grants from the National Science Foundation. Few scientists have the stature and stamina to survive like that for long, but Glenn was a master with more than a decade of steady funding.

"Hi, Bob!" he said, with his usual big smile, his thick black mustache and curly black hair making him seem younger than his years. ("It comes from drinking lots of good red wine," he'd say.) Piles of manuscripts and journals covered most of his desk and surrounding tables; banks of neatly labeled filing cabinets hinted that he had the upper hand on entropy. Glenn leaned back in his chair, hands behind his head—a characteristic gesture I'd soon come to learn. "What's up?"

I described the chirality experiments and their implications for origins research in as sexy a way as I could. Glenn nodded often, but his smile slowly faded. When I had finished, he launched into an intimidating list of his own amino acid projects already underway.

Glenn's research exploited the fact that although almost all of life's amino acids are left-handed, as soon as an organism dies, a slow, inexorable process called racemization—the random flipping of molecules from L to D and vice versa—begins. Eventually, after a few tens of thousands of years, an organism's amino acids will have completely randomized to a 50:50 mixture. This tendency for the D:L ratio of amino acids to change over time provides a powerful dating technique: The older the shell or bone, the closer its amino acids will be to a 50:50 mix. Other factors—notably the average water temperature, the acidity, and the salinity—also affect the rate of racemization; the D:L ratio in a fossil can thus provide evidence for changes in ancient environments. Glenn was one of the world's experts in determining that crucial ratio, so scientists from all over asked for his help. In one ongoing collaboration, he determined the ages of fossil eggshells from Australia, to help understand long-term changes in the continent's vegetation. Another project used clam shells to measure recent changes in the salinity of the Venetian Lagoon. He also was studying amino acids in fossil shells from the Baja Peninsula of Mexico to deduce patterns of climate change.

But his biggest and boldest effort was his long-term collaboration with Harvard paleontologist Stephen Jay Gould on the evolution of

Cerion, a beautiful little Bahamian land snail. Glenn had helped to collect countless thousands of these inch-long shells from deep pits dug into remote sand dunes. The major part of the collection was exhumed from a deserted stretch of Long Island in the Bahamas. Glenn's *Cerion* specimens displayed remarkable variations, even though all were members of a single species. Some shells were elongated, while others were almost round; some richly decorated, others almost smooth. These and several dozen other morphological characteristics provided Gould with a perfect species to test his provocative theories of evolutionary change. Glenn's job was to provide the critical dating by analyzing D:L ratios from thousands of individual shells. Once supplied with enough differently shaped shells, their ages, and the DNA analyses performed by another colleague, Gould hoped to tease out the evolutionary pathways of gradual morphological change. Years, maybe decades, of work lay ahead.

Given these commitments, Glenn was certainly too busy to take on a new project. Yet he was also intrigued. Chiral selection was a new challenge for his analytical system, and he knew a good project when he saw it.

"Looks like it's about time for lunch!" he said, abruptly changing the subject.

"Let me take you." I sensed a setup, but I would have done just about anything to secure his help.

"There's a nice little place a couple of blocks from here. Kinkead's." It wasn't a question.

"Sure, let's go."

Kinkead's specializes in seafood, to which Glenn was deathly allergic; he even had to wear protective gloves when handling his favorite *Cerion* shells. But Kinkead's had a great wine list, and Glenn had a passion for good red wines. Glenn ordered glasses of two different wines and extra glasses for each of us, so we could compare and contrast. Some months later, I learned that Kinkead's was a kind of test; had I balked at the noontime diversion, our collaboration might never have happened.

Evidently I passed. "OK," he said, and paused. "You'll have to derivatize the samples, but I'll do the analyses." So I would have to do a bit of chemical prep work, but I was in business.

Glenn had to maintain his analytical facility in the same large room as the undergraduate anthropology lab at George Washington. The first thing you notice on entering is the inordinate number of bones—dozens of human skulls, legs, ribs, and hip bones in wooden trays and glass display cases. A fully mounted human skeleton (lacking only the odd digit and forelimb) presides slack-jawed over the unsettling scene. A long, black-topped table surrounded by two dozen padded stools occupied the center of the 25×40-foot space. Glenn had a cramped 3-foot-square chemical hood on one side of the room and his arsenal of state-of-the-art gas chromatographs along the opposite wall, where any undergraduate might inadvertently bump into them. How could anyone work effectively under these conditions, I wondered? And yet one quickly learns to focus only on the diminutive vials and their secrets.

I showed up there mid-morning of the following week to prepare my amino acid samples for analysis. The aspartic acid had to be chemically modified so that the D- and L-amino acids could be separated more efficiently by gas chromatography. The amino acid molecules, which normally dry to a white powder, had to be treated so that they evaporated to a gas at high temperature. Under Glenn's guidance, I made sure each sample was completely dried down, then added a milliliter of thionyl chloride, an orange-tinged toxic liquid, and tightly capped the vials under a stream of nitrogen gas. Then we cooked the samples, two dozen at a time, in the oven.

I have never watched a scientist more meticulous in his procedures than Glenn, who proved to be one of the most exacting, finicky experimentalists I'd ever met. Like a master chef, he prepared amino acid samples for analysis the same way every time. He heated them at 100°C for one hour in a small, squat oven, instructing me to open the oven door quickly and place the tray of vials on the shelf in one swift gesture. Close the door within 4 seconds to keep the temperature at the proper level. If the temperature dropped even 0.2°C, he recorded it in his lab notebook.

An hour later, to the second, I had to remove the samples from the oven with a similar smooth motion. If I was 10 or 15 seconds late, his mustache would twitch and the discrepancy went into the notebook. One secret to Glenn's success was his absolute, rigorous reproducible procedures.

Once the rows of vials had cooled, I opened each one, dried them under a vacuum, added a second chemical (trifluoroacetic acid anhydride), sealed the vials, and heated them again for exactly five minutes. At the end of this process, each amino acid sample had been modified to a volatile form that was ready to analyze. We transferred a small volume from each into glass autosampler vials, loaded up the gas chromatograph, and set it to run overnight.

Glenn and I were paranoid about the potential for unconscious bias. We knew exactly the chiral effects we were looking for—certain faces should select L-molecules and others D-molecules, while the fracture surfaces should display no preference. So I randomly renumbered the samples and Glenn renumbered them again in his own notebook. That way, neither of us would know which sample came from which face until after we'd completed all the analyses and compared numbers. It's all too easy to see what you want to see in random data. Once the samples were prepared, we had only to wait for the automated machine to do the analyses. Glenn promised to call me the next day with our first results.

"Hi, Bob. Looks like we have some data," he reported the next afternoon. "Got a pen?" I scribbled down a long list of specimen numbers and D:L ratios. Quite a few of the numbers were close to 1.00—no effect. But there were also several values significantly higher and lower: 0.958, 1.031, 0.965, and other numbers that pointed to a possible chiral effect.

"Of course we'll have to repeat all these analyses a couple more times," Glenn added. I was to learn that performing analyses in triplicate (at a minimum) was one of his trademarks.

"What sorts of reproducibility do we have?"

"Looks like about plus-or-minus half a percent. Not bad." I was amazed. Errors smaller than 1 percent were almost unheard of in this business.

As soon as I had sorted out which analysis went with which face, a clear and compelling story began to emerge from the data. Left-handed calcite faces almost universally displayed D:L ratios a few percent less than 1.00. These faces preferentially retained L-aspartic acid. The right-handed calcite crystal faces displayed an equal and opposite affinity for D-aspartic acid. Equally important, all of the nonchiral fracture surfaces, which served as our experiment's internal control yielded D:L ratios indistinguishable from 1.00. Glenn's repeat analyses of each sample over the next week reinforced the story.

We wrote up the results quickly and submitted the short manuscript to the *Proceedings of the National Academy of Sciences*, with Hat Yoder serving as the sponsoring Academy member. The discovery that chiral crystal faces of calcite selectively adsorb D- and L-amino acids suggested not only a plausible chiral environment on the early Earth, but also a possible mechanism for making functional biological macromolecules. If adsorbed L-amino acids lined up sequentially on the crystal surface, then they might be poised to link to each other, forming a protein-like polymer of amino acids. Perhaps in this way mineral surfaces selected, organized, and assembled the first homochiral biomolecules.

As we had hoped, a few workers in the origin-of-life community noticed our results. What we had not anticipated was significant interest from chemical engineers engaged in the design and purification of chiral pharmaceuticals. Our work on chiral mineral surfaces had opened the door to a host of possible industrial applications in the $100-billion-a-year chiral drug business.

For our part, Glenn and I saw the aspartic acid study as the beginning of a long and fruitful collaboration. Next on the agenda were similar experiments with D- and L-glutamic acid, another amino acid that binds readily to calcite. We also plotted out new experiments with left- and right-handed quartz crystals. As our friendship grew, so did my interest in his other research projects, and he signed up my wife and me as field hands for his next Bahamain field season the following December.

It was during these new experiments that Glenn, uncharacteristically, began to complain of an incessant pain in his jaw. A drug-resistant tooth infection had gradually spread through his mouth and into his sinuses. Worse than the pain, the disease numbed Glenn's sense of taste. He began to lose weight rapidly. He stopped drinking wine. In March of 2002, the antibiotic Cipro seemed to turn the tide. Glenn rallied and he even agreed to visit a favorite lunch spot, Pizza Paradiso, where for the first time in weeks he managed to eat most of his lunch. We talked optimistically of our December trip to the Bahamas.

Though weakened, Glenn returned to his lab and began to recalibrate the sensitive analytical machines that had sat idle for so long. On March 27th, we enjoyed a brief, sobering visit from Steve Gould, whose magnum opus, *On the Structure of Evolutionary Theory*, had just appeared in print. Steve talked optimistically about the upcoming fieldwork, but he had just been diagnosed with a fast-spreading cancer and he tired quickly. During much of the visit, he sat in front of piles of his beloved *Cerion*, picking up one after another, pointing out unusual features. He kept saying "I need another 20 years. I just need another 20 years." [Plate 8]

But it was not to be. Stephen Jay Gould died of cancer on May 20th, fewer than two months later.

By the end of May, Glenn's infection had returned with increased virulence, spreading to his brain, confusing his thoughts. During our last halting conversation, in early October, he fretted about the long hiatus in his research. He spoke eagerly of the December field trip to the Bahamas, as if in another few weeks he'd be well again. In his delirium, he anticipated meeting Steve Gould on the island.

Glenn Goodfriend died on October 15, 2002, at the age of 51. The chance to know and work with him was one of the greatest gifts of my career, and his decline and death one of the saddest events I've ever had to experience. For months I was paralyzed by the loss. Asking anyone else to fill Glenn's shoes seemed disrespectful, like marrying again too soon after the death of a spouse. Colorful crystals lay idle in my lab. More than a hundred vials of amino acids sat unanalyzed. Only gradually, with the help of new collaborators, did the chiral-selection project get back on track.

Scientists don't know for certain—and may *never* know for certain—how life's homochirality emerged from the random prebiotic milieu, but we have targeted an expanded repertoire of promising local chiral environments. Perhaps life's molecules self-select for handedness. Or perhaps they spontaneously assemble on chiral mineral surfaces. Whatever the answer, these ideas offer years of opportunities for origin-of-life researchers (and chemical engineers, as well).

What we can say for sure at this stage is that mineral surfaces are remarkably successful at selecting, concentrating, and organizing organic compounds. Thanks to quartz, calcite, and a growing list of other crystals, the mystery of the emergence of organized molecular systems from the complex prebiotic soup seems a lot closer to being solved. It would appear that minerals played a far more central role in the origin of life than previously imagined. Armed with that understanding, chemists, biologists, and geologists are embracing a more integrated approach to one of science's oldest questions.

Interlude—Where Are the Women?

Reading your manuscript is really depressing.
Where are the women?

Sara Seager, 2004

E ven a cursory scan of this book reveals a field that has been overwhelmingly dominated by white males. Why should this be? Who's to blame?

The answer certainly isn't in the nature of the discipline. A few scientific subjects, like field geology and high-pressure research, require extraordinary physical exertion and carry a level of risk that provided a convenient excuse for decades of almost exclusive male domination. But no such hardships are associated with research on life's origins, a field that holds intrinsic fascination for men and women alike. Yet hardly a single female appeared as coauthor on any origins paper in the three decades following Stanley Miller's 1953 landmark article.

I don't know why, but I suspect that two factors played a significant role in this unfortunate, embarrassing bias.

First, the origin-of-life field is small, and by simple bad fortune several of the most prominent leaders during the 1950s through the 1970s were male professors who were at best unsupportive of women students (if not downright misogynistic). All young scientists need the encouragement of mentors and the inspiration of role models. Lacking this support system, women felt excluded from the origins club. Only within the past decade has the research environment changed enough to provide women with a more conducive environment in which to excel.

Second, the best and brightest women scientists may have shied away from the origins field in the beginning, when it held a rather dubious status in science. Everyone is interested in how life began, to be sure, but unambiguous experiments and firm conclusions are scarce. It was hard enough for many women to be taken seriously as scientists in the 1950s, 1960s, and even into the 1970s. No point in stacking the deck by entering a suspect area of research. Better to have concentrated on mainstream subjects like genetics or organic chemistry if you wanted to land a good job.

Times have changed. More than 100 women scientists participated in the 2002 Astrobiology Science Conference at NASA Ames, while female enrollments in several astrobiology PhD programs are close to 50 percent. But our community must still look in the mirror and ask why it has taken so long.

Part IV

The Emergence of
Self-Replicating Systems

Four billion years ago, life began to emerge on an Earth hellish almost beyond imagining. Volcanoes poured rivers of lava onto the land and belched noxious sulfurous vapors into the atmosphere, while meteors fell in a fitful bombardment. Violent weather and epic tides lashed the primordial coastlines. Nothing remotely lifelike graced the desolate surface.

Yet the seeds of life had been planted. The Archean Earth boasted vast repositories of serviceable organic molecules that had emerged from chemical reactions in the sunlit oceans and atmosphere, the depths of the crust, and the distant reaches of space. These molecules inevitably concentrated and assembled into vesicles, polymers, and other macromolecules of biological interest. Yet accumulations of organic macromolecules, no matter how highly selected and intricately organized, are not alive unless they also possess the ability to reproduce.

Most experts agree that life can be defined as a chemical phenomenon possessing three fundamental attributes: the ability to grow, the ability to reproduce, and the ability to evolve. The first of these three characteristics, individual cell growth, has occurred on an all but invisible scale for most of Earth's history; for 3 billion years—from about 3.8 billion years ago to about 700 million years ago—the largest living organisms on Earth were microscopic single-celled objects. The third characteristic, evolution, proved to be slow and subtle: It took billions of years for cells to learn some of the most familiar biochemical tasks, such as using sunlight for energy or oxygen for respiration. But reproduction, life's most dynamic overriding imperative, must have operated with an inexorable, geometric swiftness. In the geological blink of an eye, rapidly reproducing populations of cells doubled and redoubled, spreading through Earth's oceans, colonizing and transforming the globe. Reproduction, more than any other characteristic of life, set the world of life apart from the prebiotic era.

For origin-of-life researchers, creating a self-replicating molecular system in a test tube has become the experimental Holy Grail. It has proved vastly more difficult to devise an experiment to study chemical self-replication than to simulate the earlier stages of life's emergence—prebiotic synthesis of biomolecules, or the selective concentration and organization of those molecules into membranes and polymers, for example. In a reproducing chemical system, one small group of molecules must multiply again and again at the expense of other molecules, which serve as food. It's an extraordinarily difficult chemical feat, but the rewards for success are immense. Imagine being the first scientist to create a lifelike chemical system in the laboratory!

14

Wheels Within Wheels

The origin of metabolism is the next great virgin territory
which is waiting for experimental chemists to explore.
Freeman Dyson, *Origins of Life*, 1985

Scientists who study life's origins divide into many camps. One group of researchers claims that life is a cosmic imperative that emerges anywhere in the universe if conditions are appropriate, while their rivals view life on Earth as a chance (and probably unique) event. One camp claims that some sort of enclosing membrane must have preceded life; another counters with models of "flat" surface life. Some researchers insist that life's origin depended on solar energy, while others point to Earth's internal heat as a more likely triggering source. Lacking adequate observational or experimental evidence to arbitrate these divergent views, positions sometimes become polarized and inflexible.

Perhaps the most fundamental of the scientific debates on origin events relates to the timing of two essential biological processes, metabolism and genetics. Metabolism is the ability to manufacture biomolecular structures from a source of energy (such as sunlight) and matter scavenged from the surroundings (usually in the form of small molecules). An organism can't survive and grow without an adequate supply of energy and matter. Genetics, by contrast, deals with the transfer of biological information from one generation to the next—a blueprint for life via the mechanisms of information-rich polymers, such as DNA and RNA. An organism can't reproduce without a reliable means to pass on its genetic information.

The problem as it relates to origins is that metabolism and genetics constitute two separate, chemically distinct systems in today's cells, much as your circulatory and nervous systems are physically distinct networks in your body. Nevertheless, just like your blood and your brains, metabolism and genetics are inextricably linked in modern life. DNA holds genetic instructions to make hundreds of molecules essential to metabolism, while metabolism provides both the energy and the basic building blocks to make DNA and other genetic materials. Like the dilemma of the chicken and the egg, it's difficult to imagine a time when metabolism and genetics were not intertwined. Consequently, origin-of-life researchers engage in an intense ongoing debate about whether these two aspects of life arose simultaneously or independently and, if the latter, which one came first.

Most experts seem to agree that the simultaneous emergence of metabolism and genetics is unlikely. The chemical processes are just too different, and they rely on completely different sets of molecules. It's much easier to imagine life arising one small step at a time, but what is the sequence of emergent steps?

Those who favor genetics as the first step argue in part on the basis of life's incredible complexity. They point to the astonishing intricacy of even the simplest living cell. Without a genetic mechanism, there would be no way to ensure the faithful reproduction of all that complexity. Metabolism without genetics, they say, is nothing more than a bunch of overactive chemicals. Biologists, who must deal with complex cells and whose academic curriculum is dominated by the powerful, unifying spell of the genetic code, seem inclined to adopt this view without reservation.

Other scientists, myself included, tend to be more influenced by what we perceive as the underlying chemical simplicity of primitive metabolism. We are persuaded by the principle that life emerged through stages of increasing complexity. Metabolic chemistry, at its core, is vastly simpler than genetics, because it requires relatively few small molecules working in concert to duplicate themselves. Harold Morowitz and Günter Wächtershäuser, among others, agree that the core metabolic cycle—the citric acid cycle, which lies at the heart of every modern cell's metabolic processes—survives as a chemical fossil almost from life's beginning. This comparatively simple chemical cycle is an engine that can bootstrap all of life's biochemistry, including the key molecules of genetics.

We conclude that if life arose through a sequence of ever more complex emergent steps, then bare-bones metabolism seems the more likely precursor. Such a no-frills metabolic system, furthermore, might easily be enclosed in a primitive membrane of the type David Deamer makes in his lab. My sense is that many chemists, physicists, and geologists (not many of whom deal with DNA on a regular basis) are persuaded by this metabolism-first point of view.

Origin-of-life scientists aren't shy about voicing their opinions on the metabolism-first versus genetics-first problem, which will probably remain one of the hottest controversies in the field for some time. Meanwhile, as this debate fuels animated discussions at conferences and in print, several groups of researchers are attempting to shed light on the issue by devising self-replicating chemical systems—metabolism in a test tube.

SELF-REPLICATING MOLECULES

The simplest imaginable self-replicating system consists of a single molecule that makes copies of itself. In the right chemical environment, such an isolated molecule will become two, then four, then eight molecules and so on in a geometrical expansion. Such a molecule is autocatalytic—that is, it acts as a template that attracts and assembles its own components from an appropriate chemical broth. Single self-replicating molecules are intrinsically complex in structure, but or-

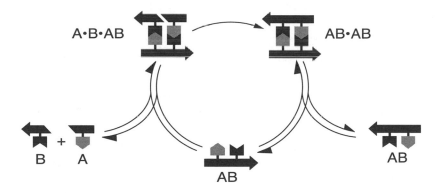

Self-replicating molecules catalyze their own formation from smaller building blocks. In this example, one molecule AB plus individual A and B molecules combine to make new AB molecules.

ganic chemists have managed to devise several varieties of these curious beasts.

Self-replicating molecules possess a distinctive feature known as self-complementarity. Pairs of complementary molecules—molecules of different species that fit snugly together because of their shape and the arrangement of their chemical bonds—turn out to be relatively common in nature. Many of the proteins in your body, for example, function because they are complementary to food, neurotransmitters, toxins, or other molecules of biological importance. For this very reason, the synthesis of complementary molecules has become a central task of computer-aided drug design. If you can design a molecule that fits into pain receptors and thereby blocks pain signals, you can make a lot of money. Only a handful of known molecules are *self*-complementary, however, and even fewer can self-replicate.

Hungarian-born chemist Julius Rebek, Jr., now at the Scripps Research Institute in La Jolla, California, has spent many years designing such molecules. In 1989 Rebek's group employed three smaller molecules: the purine adenine; the two-ring cyclic molecule napthalene; and imide, a nitrogen-containing compound. When bonded end-to-end, the resulting adenine–napthalene–imide molecule is self-complementary and, if immersed in a solution containing lots of these three individual molecules, will make copies of itself.

Reza Ghadiri, leader of another productive research group at the Scripps Institute, focuses on yet another chemical system—self-replicating peptides, which are arguably much more relevant than most other molecules when it comes to the origin of life. Peptides, like proteins, are chainlike molecules built from long sequences of amino acids; their constituents are drawn from the same library of 20 amino acids that make up proteins, but they are typically dozens instead of hundreds of amino acids long. In 1996, Ghadiri's group reported the first synthesis of a self-replicating peptide—a sequence of 32 amino acids. For self-replication to occur, however, this peptide had to be "fed" two reactive fragments, one of them 15 and the other 17 amino acids long. Given a steady supply of these two precursor fragments, new peptides formed spontaneously.

Self-complementary strands of the genetic molecule DNA display similar self-replicating behavior. As James Watson and Francis Crick famously discovered in 1953, the classic double-helix structure of the DNA molecule is like a long, twisted ladder, whose vertical supports are chains of alternating phosphate and sugar molecules and whose rungs consist of the complementary pairs of molecules called bases: Cytosine (C) always pairs with guanine (G), while adenine (A) always pairs with thymine (T). Consequently, a single DNA strand with the base sequence ACGTTTCCA, say, is complementary to a second single

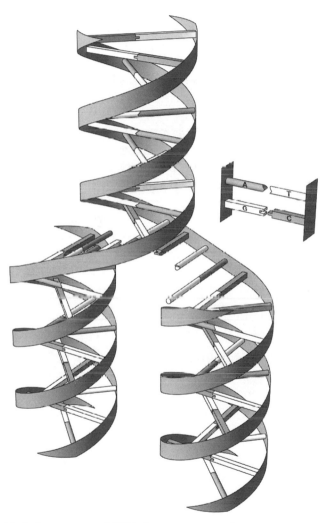

The double-helix structure of DNA features two complementary sequences of the bases A, C, G, and T. A always binds to T, while C always binds to G. The molecule replicates by splitting down the middle and adding new bases to each side. Thus, one strand becomes two.

strand TGCAAAGGT. When the two strands separate, each can then make a copy of the original double helix—as noted in one of the most celebrated and oft-quoted sentences in the history of biology, from Watson and Crick's landmark paper: "It has not escaped our attention that the specific pairing we have postulated immediately suggests a possible copying mechanism for the genetic material."

The vast majority of DNA strands are not self-complementary, because the two halves of the double helix have different base sequences. But in 1986, German chemist Guenter von Kiedrowski and co-workers synthesized the first of the so-called "palindromic," self-complementary DNA strands—the six-nucleotide sequence CCGCGG, which makes exact copies of itself if sufficient supplies of fresh C and G are provided.

It seems almost magical for a molecule to make copies of itself. Nevertheless, these self-replicating macromolecules do not meet the minimum requirements for life on at least two counts. First, such systems require a steady input of smaller highly specialized molecules—synthetic chemicals that must be supplied to the system from somewhere. Under no plausible natural environment could sufficient numbers of these component molecules have arisen independently. Furthermore—and this is a key point in distinguishing life from non-life—self-replicating molecules do not change and evolve, any more than a photocopy can evolve from an original.

SELF-REPLICATING MOLECULAR SYSTEMS

More relevant to metabolism are systems of two or more molecules that form a self-replicating cycle or network. Such systems are now the subject of intense research, and a variety of strategies for molecular self-replication have been identified. In the simplest so-called "cross-catalytic" system, two molecules (call them AA and BB) form from smaller feedstock molecules A and B. If AA catalyzes the formation of BB, and BB in turn catalyzes the formation of AA, then the system will sustain itself as long as researchers maintain a reliable supply of food molecules A and B.

It's easy for theorists to elaborate on such a model. Rather than two cross-catalytic molecules, imagine a system with 10 or 20 molecules, each of which promotes the production of another species in the system. Santa Fe Institute theorist Stuart Kauffman points to such

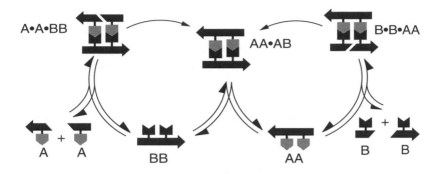

In a cross-catalytic system, two different molecules (in this example designated AA and BB) catalyze the formation of each other. More complicated cross-catalytic systems with numerous molecules may have been the first self-replicating cycle on the early Earth.

"autocatalytic networks" as the most likely form of metabolic protolife, since that's exactly what modern life does. Given the complexity of living cells, he proposes that the molecular repertoire of the first autocatalytic network might have to be expanded to as many as 20,000 different interacting molecular species. Accordingly, Kauffman drafts complex spaghetti-like illustrations of hypothetical reactants, each playing a key role in the integrated network.

Kauffman sees an important payoff in the behavior of such elaborate catalytic networks. Unlike a single self-replicating molecule, such vast networks of reactive molecules have the potential to increase in complexity and efficiency through numerous different interactions among molecules. The large number and variety of molecular interactions foster the emergence of new metabolic pathways and nested cycles of synthesis. Like a living metabolic cycle, Kauffman's networks incorporate a certain degree of sloppiness. Through such variations, the system can change by shifting to new, more efficient reaction pathways and in the process display a defining trait of life—the ability to evolve.

THE GAME OF METABOLISM

Kauffman's theory provides an intuitively appealing model for the emergence of primordial metabolic life. For one thing, if molecules in a complex autocatalytic system efficiently catalyze each other, then there's no need, at least at first, for a genetic mechanism. Accurate replication of the entire system is, in effect, intrinsic to the network. At the same time, the system has the potential to vary and gradually evolve. Growth, reproduction, evolution—the autocatalytic network would appear to meet the minimum criteria for life.

Kauffman's model sounds good on paper, but there's a dramatic gap between plausible theory and actual experiment. For example, what exactly is the chemical "A"? What is "B"? When asked, Kauffman shrugs and says, "That's for the chemists to figure out." And that's what chemists have done.

If you want to make protolife in the lab, it's best to start by designing a simple metabolic cycle. Metabolism is a special kind of cyclical chemical process with two essential inputs. First you need a source of energy—preferably chemical energy, since that's what all known simple living cells use. (Photosynthetic cells use sunlight, to be sure, but these advanced organisms possess layers of chemical complexity far beyond those of more ancient, simple cells.) Second, you need reliable supplies of simple molecules, such as carbon dioxide and water, to provide the raw materials.

Living cells undergo chemical reactions not unlike burning, in which two chemicals (oxygen and some carbon-rich fuel) react and release energy. However, the objective in metabolism, unlike in an open fire, is to capture part of that released energy to make new useful molecules that reinforce the cycle. So metabolism requires a sequence of chemical reactions that work in concert.

Three basic rules govern this game of metabolism:

Rule 1—The Cycle: Describe a cycle consisting of a sequence of progressively larger molecules. The largest of these molecules should then be able to split into two smaller molecules, both of which are also in the cycle. In this way you end up with two sets of molecules, where before there was only one set and the cycle keeps on reproducing itself.

Rule 2—The Environment: Identify a plausible prebiotic environment that provides a reliable supply of raw materials (i.e., small molecules made up of carbon, nitrogen, sulfur, and other essential elements) and energy (preferably the chemical energy of unstable minerals). Furthermore, *all* of the different types of molecules in the cycle have to be able to survive in that environment long enough to keep the cycle going. Extremes in temperature, pressure, or acidity, for example, may stabilize some kinds of molecules, while destroying others.

Rule 3—Continuity: An unbroken chemical history must link Earth's earliest metabolic cycle with the metabolism of today's cells. Deep within the metabolism of all of us, there are likely to be "fossil" biochemical pathways that point to life's simpler beginnings. So any plausible model must conform to this "principle of continuity" and use molecules and energy sources that lie close to the basic metabolism of modern organisms.

Experimental evidence provides the ultimate test of any chemical theory. Each chemical step in a hypothetical metabolic cycle is a potential experiment, so you get bonus points if these crucial experiments work.

Three influential models exemplify the metabolism game at its best: the "Protenoid World" of Sidney Fox, the "Thioester World" of Christian de Duve, and a group of related hypotheses based on the reverse citric acid cycle, including the elaborate "Iron–Sulfur World" hypothesis of Günter Wächtershäuser.

THE PROTENOID WORLD

The granddaddy of all metabolism-first models emerged as the brainchild of protein chemist Sidney Fox, who began thinking about origin-of-life chemistry shortly after the publication of Stanley Miller's 1953 paper. Fox's career was strongly influenced by Thomas Hunt Morgan, a famed geneticist and a member of Fox's Caltech thesis committee. "Fox," Morgan would often remark, "all the important problems of life are problems of proteins." Fox took this lesson to heart and began studying proteins' role in life's origin as a faculty member at Florida State University in the mid-1950s.

He imagined a hot primordial Earth where amino acids from the soup dried and baked on cooling volcanic rocks. He mimicked these conditions in his lab, drying amino acids on a hot surface at 170°C, and found that his chemicals quickly polymerized into a lumpy substance he called "protenoid." This discovery, announced in 1958, would shape his checkered three-decade career in origins research.

Fox's protenoids displayed fascinating behavior. They acted as catalysts, though to a rather modest degree compared to the modern protein catalysts that promote chemical reactions in every cell. When placed in hot water, they spontaneously turned into microbe-sized spheres, and sometimes these divided. He claimed that these spheres, though not as orderly as biological proteins, possessed a nonrandom structure with a permeable bilayer membrane. Increasingly, Fox suggested that the protenoids were "lifelike" and represented the key to life's origins.

Fox and a small army of students carried out experiments to bolster his origin model. Solutions of protenoids were heated and cooled in a cycle that produced structures he called "microspheres"—cell-like entities that could absorb more protenoids, grow, and divide, thus forming a second generation of microspheres. These self-replicating objects, growing from nutrients in solution and lacking any genetic mechanism, formed the basis of a true metabolism-first model.

At first blush, the Protenoid World might seem to score well in the game of metabolism. Rule 1: The cycle is formed from widely available amino acids, which readily form protenoids that grow and divide. Rule 2: The environment, a volcanic setting near a tidal zone, is realistic for the early Earth. Rule 3: Protenoids seemed to conform to the principle of biological continuity, because closely related proteins are fundamental to modern life. What's more, experiments seemed to support each step of the hypothesis.

For a time, Fox's career thrived. Starting in 1960, he received generous grants from NASA's Exobiology program, which supported a variety of research projects on the origin and evolution of life. Fox's funds soon exceeded $1.5 million and allowed him to establish his own Institute of Molecular Evolution at the University of Miami in 1964. A steady stream of graduate students investigated protenoid properties.

As Sidney Fox increasingly fell in love with his Protenoid World model, his claims became more and more extreme. The origin-of-life problem had been solved, he said, and protenoids are "alive in some primitive way." In his 1988 book, *The Emergence of Life*, he even made the bizarre and unsubstantiated claim that his protenoid microspheres possessed a kind of "rudimentary consciousness."

As early as 1959, the mainstream origin-of-life community, led principally by Stanley Miller and Harold Urey, had begun distancing themselves from Fox's claims. They resented what they regarded as the sensationalistic use of terms such as "protenoid" and "lifelike." They scoffed at the idea that amino acids might have baked into anything useful on a bed of lava. They questioned whether the protenoid structures were truly nonrandom in structure. The purported lifelike behaviors of microspheres were also challenged, especially the idea that they "replicated" in the biological sense. Equally strong was their objection that the Protenoid World scenario ignored the knotty problem of genetics. How would DNA and RNA have arisen in such a world?

Despite myriad objections to the theory, and the increasing scientific isolation of Fox, the Protenoid World was influential for at least two reasons. Not only did Fox develop the first comprehensive metabolism-first model, but he also championed the important philosophical position that origin processes might be nonrandom and deterministic—a view later embraced by numerous other origins workers. Nevertheless, by the late 1970s, Sidney Fox's efforts had been marginalized, and they remained the target of jokes and derision until his death in 1998.

THE THIOESTER WORLD

The Belgian chemist Christian de Duve made his scientific name studying the structural and functional organization of cells—work that won him the Nobel Prize in physiology or medicine in 1974. In an oft-repeated pattern, he then turned his attention to the origin of life, tackling the classic question from his perspective as a cell biologist. Not surprisingly, that background influences his origins theory.

De Duve takes a rather ambiguous stand in the metabolism- versus genetics-first controversy. For him, the simplest imaginable living thing is a cell-like entity that has a full complement of genetic material to control cellular functions. But he also recognizes the futility of jump-starting genetics without a well-established, elaborate arsenal of chemical reactions—what he calls "protometabolism" (as opposed to the more complex metabolism of modern cells). It's really a question of where you place the boundary between nonliving and living systems. I've argued that the complex sequence of emergent events leading from geochemistry to life requires a rich taxonomy of intermediate states. De Duve appears to agree: His "Age of Chemistry" (protometabolism) precedes the "Age of Information" (genetics), which in turn precedes the "Age of the Protocell" (genetics and metabolism combined). Only then, with the merger of metabolism and genetics, does modern life truly emerge.

Let's assume, de Duve says, that some early Earth environment (rather vaguely specified as a "volcanic setting" in his writings) was rich in hydrogen sulfide gas and iron sulfide minerals, along with the usual cast of carbon-based molecules. One chemical consequence of this sketchy scenario might be the production of sulfur-containing molecules called thioesters, which incorporate a strong bond between one sulfur atom and one carbon atom—a bond that can release a lot of energy if broken. De Duve's hypothesis rests on the assumption that a steady supply of these energy-rich thioester molecules was available on the ancient Earth—hence the Thioester World.

Why focus on thioesters? For one thing, they are essential in the metabolism of modern cells. By breaking the strong carbon–sulfur bond, they can transfer energy to promote metabolic reactions, thus building larger molecules from smaller ones. In the Thioester World, a steady supply of thioesters provides the chemical energy required to drive protometabolism.

Of particular import, thioesters have a propensity to bond with amino acids, which would have been readily available in the prebiotic soup. Remarkably, when placed in solution, these amino acid–thioester groups spontaneously assemble into peptidelike chains (de Duve calls them "multimers," because they differ in some chemical details from true peptides). At first, the multimers thus formed would have had little effect on the chemical mix. But, he speculates, eventually some of these big molecules by chance acquired catalytic properties (reminiscent of Fox's protenoids). Gradually, as the environment became enriched in chemically active multimers, an autocatalytic cycle might have emerged, producing, he says, "protoenzymes required for protometabolism."

The Thioester World hypothesis fits nicely with the primordial-soup paradigm favored by Miller and his allies. De Duve's scenario relies on the environment to provide a steady input of various molecular building blocks, including amino acids, carbohydrates, and of course the essential energy-rich thioesters themselves. Such a model metabolism, in which the first cells eat and assemble components from their surroundings, is said to be heterotrophic (from the Greek, "other nourishment"). Heterotrophs must scavenge their molecules from the environment.

So how does de Duve's Thioester World score in the game of metabolism? He certainly gets high marks for identifying a viable prebiotic environment rich in plausible simple molecules. Using thioesters as a reliable, renewable energy source is especially attractive, because it echoes the action of thioesters in modern biology. But he never specifies the molecules that comprise his metabolic cycle; the chemistry is vague. For de Duve, protometabolism is an awkward, transient, ill-defined phase that is necessary to set the stage for genetics. Thioesters promote the synthesis of catalytic multimers, which in turn promote the manufacture of a genetic molecule such as RNA. Only then does a well-defined biochemistry take root.

The bottom line: de Duve's scenario is appealing in its broad outlines, but lacking in specific chemical details. And so, if you like details, there's no better place to turn than Günter Wächtershäuser's Iron–Sulfur World.

15

The Iron–Sulfur World

You don't mind if I brag a little, but something like this has
never been done in the entire field.
                                    Günter Wächtershäuser, 1998

In bold concept, epic sweep, and sheer mass of detailed predictions, Günter Wächtershäuser's Iron–Sulfur World hypothesis for the origin of life stands alone. Since its first presentation in 1988, this theory has changed the landscape of origins research by calling into question many of the most deeply rooted assumptions about life's beginnings.

Wächtershäuser (whom we first met in Chapter 8 because of his pleasing reliance on minerals) argues for a strikingly original metabolism-first model, based on an autotrophic (self-nourishment) model. Autotrophic metabolism begins with chemical simplicity. All of life's essential biomolecules are manufactured in place, as needed, from the smallest of building blocks: carbon dioxide (CO_2), water (H_2O), ammonia (NH_3), hydrogen sulfide (H_2S), and so forth. All chemical synthesis is accomplished stepwise, just a few atoms at a time. [Plate 8]

The contrast between heterotrophic and autotrophic existence is profound and represents a fundamental point of disagreement among origin-of-life researchers. Supporters of a heterotrophic origin argue that it's much easier for a primitive cell to use the diverse molecular products available in the prebiotic soup rather than to make them from scratch. Why go to the trouble of synthesizing lots of amino acids if they're already available in the environment? Modern heterotrophic cells are much simpler than autotrophic cells, because they don't need all the complex chemical machinery to manufacture amino acids, carbohydrates, lipids, and so forth. It makes sense to assume that the simpler mechanism—heterotrophy—came first.

Autotrophic advocates are equally insistent that true simplicity lies in building molecules a few atoms at a time, with just a few basic kinds of chemical reactions. Furthermore, such a mechanism is philosophically attractive: Autotrophism is deterministic. Rather than depending on the idiosyncrasies of a local environment for biomolecular components, autotrophic organisms make them from scratch the same way every time, on any viable planet or moon, in a predictable chemical path. For hardcore advocates of autotrophy, the random, hodgepodge, hopelessly dilute prebiotic soup is irrelevant to the origin of life. With autotrophy, biochemistry is hardwired into the universe. The self-made cell emerges from geochemistry as inevitably as basalt or granite.

BACK TO THE VENTS

As with any metabolic strategy, the Iron–Sulfur World model requires a source of energy, a source of molecules, and a self-replicating cycle. Here's how Wächtershäuser does it.

His model relies on the abundant chemical energy of minerals out of equilibrium with their environment. He begins by suggesting that the common iron sulfide mineral pyrrhotite (FeS), which is deposited in abundance at the mouths of many hydrothermal vents, is unstable with respect to the surrounding seawater—that is, given time, it will spontaneously decompose to other more stable chemicals. As a consequence, iron sulfide combines with the volcanic gas hydrogen sulfide (H_2S) to produce the shiny mineral pyrite (FeS_2) plus hydrogen gas (H_2) and a jolt of energy:

$$FeS + H_2S \rightarrow FeS_2 + H_2 + Energy$$

Given that energetic boost, hydrogen reacts immediately with the carbon dioxide (CO_2) in seawater to synthesize organic molecules such as formic acid ($HCOOH$).

$$Energy + H_2 + CO_2 \rightarrow HCOOH$$

Wächtershäuser envisions cascades of these reactions coupled to build up essential organic molecules from CO_2 and other simple gases. When these speculations were first proposed, precious little evidence backed them up, but a flurry of experiments in the past decade and a half has lent support to the Iron–Sulfur World hypothesis. Wächtershäuser instigated the initial tests, joining with a team of German scientists from the University of Regensburg to study the reaction of iron sulfide and hydrogen sulfide. Sure enough, just as predicted, they produced pyrite and hydrogen gas. Their brief 1990 paper, "Pyrite formation linked with hydrogen evolution under anaerobic conditions," appeared in *Nature*.

The Iron–Sulfur World hypothesis also makes the unambiguous prediction that iron sulfide minerals promote a variety of organic reactions. In 1996, the Dutch scientists Wolfgang Heinen and Anne Marie Lauwers simulated Wächtershäuser's Iron–Sulfur World scenario by studying reactions of powdered iron sulfide and hydrogen sulfide in water with an atmosphere of carbon dioxide. As predicted, they found the synthesis of organic compounds.

Subsequent experiments by Wächtershäuser and colleagues in Germany used iron, nickel, and cobalt sulfides to synthesize acetate, an essential metabolic molecule with two carbon atoms that plays a central role in countless biochemical processes. They expanded on this success by adding amino acids to their experiments and making peptides—yet another essential step to life.

Our research group in Washington also got into the act. In one set of gold-capsule experiments with iron sulfide at high temperature and pressure, we succeeded in producing a variety of organic molecules, including the key metabolic compound pyruvate. We were especially interested to learn whether other minerals could perform the same task. Hydrothermal zones typically feature a variety of minerals in addition to the common iron sulfides; compounds of nickel, cobalt, zinc, copper, and other metals abound. So our Carnegie Institution team ran more than 300 gold-capsule experiments with a dozen different minerals at high pressure and temperature and found organic synthesis reactions in almost every run.

These experiments have led to an unambiguous conclusion: Common sulfide minerals can promote a variety of interesting synthesis reactions. That's good news for Wächtershäuser's model, but life requires more than a random assortment of chemical reactions—it requires a metabolic cycle. What self-replicating cycle of reactions led to the Iron–Sulfur World?

THE REVERSE CITRIC ACID CYCLE

A centerpiece of the Iron–Sulfur World hypothesis is Wächtershäuser's conviction that life began with a simple, self-replicating cycle of compounds similar to the one that lies at the heart of every modern cell's metabolism—the reverse citric acid cycle. A growing cadre of scientists concurs, but none has been more outspoken or articulate than Harold Morowitz, my colleague at George Mason University.

For the better part of two decades, Morowitz has preached a simple philosophy: If you want to understand the chemical emergence of life, look to life's most basic chemistry—the biochemical pathways shared by all organisms. In the case of metabolism, that core biochemistry is found in the citric acid cycle, also known to generations of high-school biology students as the Krebs cycle or the TCA cycle. This circular sequence of 11 small molecules, each made of carbon, hydrogen, and oxygen, forms the core metabolism in every cell. The cycle starts with the 6-carbon molecule citric acid, which is gradually pared down, a few atoms at a time, to make a succession of smaller molecules— 5-carbon malate, 4-carbon oxaloacetate, 3-carbon pyruvate, and so forth—each of which is the starting material for all sorts of other essential biomolecules. Add ammonia to pyruvate and you get the amino acid alanine; add carbon dioxide to acetate and you get the building blocks of lipids. Animals and plants use this cycle as the starting point for synthesizing just about everything biochemical. For many modern organisms, the citric acid cycle is an engine of biosynthesis.

In the mid-1960s, biologists discovered that some primitive microbes run the citric acid cycle the "wrong way" round—a pathway that can be called the "reverse" citric acid cycle. Starting with the 2-carbon molecule acetate (one of the newsworthy compounds synthesized in Wächtershäuser's experiments), the cell adds a carbon dioxide molecule to make 3-carbon pyruvate. Then add another carbon dioxide molecule to make 4-carbon oxaloacetate, and so on. Around the cycle we go, until we get to 6-carbon citric acid. At this point the cycle does something surprising. Citric acid splits into one acetate molecule and one oxaloacetate molecule. Suddenly we have the building blocks for two cycles, so around we go again. The reverse citric acid cycle is a self-replicating synthesis engine that doubles on every turn. In this way, a microbe can build all of its essential biomolecules from the simplest of building blocks—water and gas.

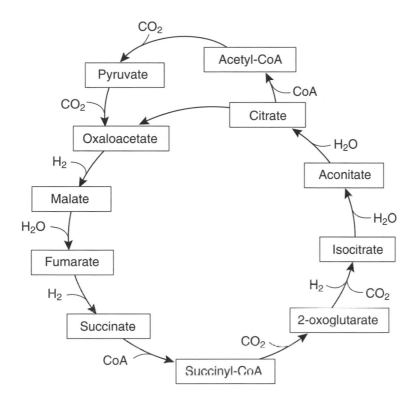

The reverse citric acid cycle is an engine of synthesis that may have been life's first metabolic cycle.

Harold loves to lecture on the foundations of metabolism, and he does so with simplicity and elegance. At a recent seminar, he used no aids but a few hand-scrawled viewgraphs and a big chart of all known metabolic pathways (he calls it "my security blanket"). At the heart of that intricate chart lies the simple citric acid cycle—the core chemistry of life itself. "We're looking for a fundamental theory of biology," he says. "When no one is listening, we call it the Grand Unified Theory of Biology."

Harold lectured to the packed, attentive room sitting back in an easy chair, his arms sweeping in wide circles. He reiterated the basic concept of the reverse citric acid cycle: Start with two small molecules—acetate with two carbon atoms and oxaloacetate with four carbon atoms. Then start tacking on carbon dioxide and water—a carbon atom here, an oxygen atom there, a few more hydrogen atoms—until, after just a few steps, you wind up with two citric acid molecules. These citric acid molecules split to begin new cycles. The various intermediate molecules of the cycle serve as starting points for the synthesis of all other key biomolecules, including amino acids, sugars, and lipids. Today, cycles of synthesis lie within cycles, which in turn lie within more cycles, but at the very core is the reverse citric acid cycle. "It's the way CO_2 was incorporated into biology before photosynthesis," explains Harold.

It's all well and good to point to today's biochemistry as a clue to life's origin, but there's a key difference between modern cells and the presumed first metabolic cycle. Today, enzymes—protein catalysts of incredible complexity—promote each of the 10 separate chemical reactions in the "simple" reverse citric acid cycle. One of these catalytic enzymes speeds up the split of citric acid into acetate and oxaloacetate. Another facilitates the addition of carbon dioxide to pyruvate to make oxaloacetate. Without these enzymes to increase chemical efficiency, today's cells wouldn't stand a chance in the competitive struggle for resources.

How did the first metabolic cycle go it alone, without an enzymatic boost? Some of the requisite steps, such as combining carbon dioxide and pyruvate to make oxaloacetate, are energetically unlikely. The trouble is that if one step fails, then the whole cycle fails. One of the clever proposals in Wächtershäuser's model is that sulfide minerals promote simple metabolic reactions. It just so happens that many modern metabolic enzymes have at their core a small cluster of iron or nickel and sulfur atoms—clusters that look exactly like tiny bits of sulfide minerals. Perhaps ancient minerals played the same role as modern enzymes.

Wächtershäuser invokes another chemical trick based on sulfur. In his model, the very first cycle was helped along by substituting hydrogen sulfide for water (H_2S for H_2O) in several crucial reactions, thus forming sulfur-bearing analogs of citric acid cycle compounds. He predicts that this simple chemical substitution leads to faster, energetically more favorable reactions. What's more, reactions with sulfide naturally couple to the formation of pyrite—the physical foundation of the Iron–Sulfur World. Once a sulfide version of the reverse citric acid cycle was well established on the primitive Earth, it was only a matter of time before it switched from hydrogen sulfide (which is common only at volcanic vents) to water (which is everywhere).

EXPERIMENTS

What do experiments have to say about the reverse citric acid cycle? In the late 1990s, our Carnegie Institution group initiated a series of gold-tube experiments to see how citric acid degrades in hot, pressurized water reminiscent of conditions at hydrothermal vents. Ideally, we would have studied the system with hydrogen sulfide gas, to match the details of Wächtershäuser's model, but H_2S is nasty stuff, technically difficult to load, and potentially deadly if a gas cylinder leaks. We stuck to the closely related water-based system.

By this time, we had our protocols down pat. I loaded the capsules with a citric acid solution, Hat Yoder set up the high-pressure runs (typically at 2,000 atmospheres and 200°C), and George Cody analyzed the runs. George soon recognized two distinct reaction pathways. Some citric acid molecules followed what he called the alpha pathway, which begins with citric acid splitting into acetate plus oxaloacetate, just like the reverse citric acid cycle is supposed to work. But there the similarity ended. Oxaloacetate rapidly decomposed to pyruvate plus carbon dioxide, and much of the pyruvate then decomposed to acetate. Rather than build up new useful molecules, we destroyed the ones we had.

The next logical step was to take pyruvate and add sulfide minerals and carbon dioxide—essential ingredients in the Iron–Sulfur World. But try as we might, no combination of minerals and reactants would promote the essential reaction from pyruvate, a 3-carbon molecule, to oxaloacetate, a 4-carbon molecule. Without that step, we had to conclude that the alpha pathway is a dead end that destroys citric acid. Oxaloacetate is the real stopper. Under no plausible prebiotic conditions have we found a way to make oxaloacetate from pyruvate plus carbon dioxide. Pressure doesn't seem to help, nor does any realistic combination of mineral catalysts. Even at modest hydrothermal temperatures less than 100°C, oxaloacetate breaks down rapidly in water.

George didn't give up. Other citric acid molecules had followed what he called the beta pathway, which begins in more or less the normal way, with citric acid giving up a water molecule to make aconitate, another 6-carbon compound. Then a series of reactions that release

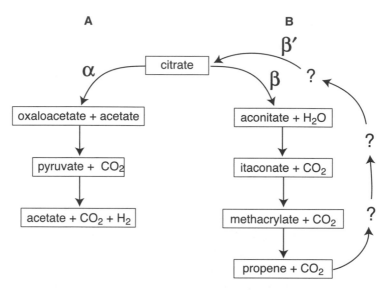

George Cody identified two distinct reaction pathways in experiments on citric acid at high temperature and pressure. (A) Some citric acid molecules followed what he called the alpha pathway, which begins with citric acid splitting into acetate plus oxaloacetate, which then fragments in a nonbiological manner. (B) Other citric acid molecules followed the beta pathway, which begins with citric acid giving up a water molecule to make aconitate, another 6-carbon compound. Then a series of reactions that release carbon dioxide produced compounds with five, then four, then three carbon atoms in succession. Cody found evidence that this pathway might be reversed to complete a ?? metabolic loop.

carbon dioxide produced compounds with five, then four, then three carbon atoms in succession.

Is this reaction series another dead end for citric acid? Or might the beta pathway be part of an alternative self-replicating citric acid cycle? In subsequent experiments, Cody found evidence that some steps of the beta pathway can be reversed in the presence of nickel sulfide, a common hydrothermal vent mineral. Perhaps an abiotic route from carbon dioxide to citric acid exists after all. We're still some way from demonstrating a self-replicating citric acid cycle, but we haven't given up yet.

And perhaps someday some brave soul will try these experiments all over again with hydrogen sulfide—the true test of Wächtershäuser's model.

FLAT LIFE

Supposing Wächtershäuser is right, what did the Iron–Sulfur World look like?

Recall that cellular life requires a membrane—a chemical barrier to separate the organized insides from the chaotic outsides. Accordingly, most origin-of-life researchers accept the idea of a cell-like lipid vesicle that surrounded the first self-replicating metabolic cycle. But an enclosing lipid membrane may not be the only possibility.

Glasgow-based geoscientists Michael Russell and Allan Hall, both experts in sulfide ore deposits, focus on probable geochemical environments in their clever "iron sulphide bubble" scenario. Unlike Wächtershäuser, they do not rely on the reactive potential of iron sulfides to drive metabolism, nor do their sulfides serve as a solid footing for life. Rather, in their scenario iron sulfide bubbles form spontaneously at hydrothermal vents, where the less acidic vent water contacts the more acidic ocean water. These bubbles form primitive cell-like structures that enclose metabolic chemicals. The bubbles also maintain a strong contrast of acidity between inside and outside—an energetic difference that can promote metabolic reactions.

By contrast, Wächtershäuser advocates "flat life." The first self-replicating entity in his proposed Iron–Sulfur World was, as we saw in Chapter 8, a thin layer of chemical reactants on a sulfide mineral surface. The entity grew laterally, spreading from mineral grain to mineral grain as an invisibly thin organic film. Bits of these layers could break off and reattach to other rocks, like cloned colonies. Given time, different minerals and environmental conditions might have induced variations in the film, fostering new "species" of flat life.

The bold, heretical concept of flat life—a self-replicating chemical layer of molecules built on a solid mineral foundation—raises an intriguing geochemical possibility. A simple layered collection of molecules might be more tolerant of high temperatures and other environmental extremes than life based on nucleic acids, which break down close to 100°C. If so, then colonies of flat life might exist today in deep zones of Earth's crust. Such film-like molecular systems might persist for eons, because they survive at extreme conditions beyond the predation of more efficient cellular life.

If so, how would we know? Such a layer would be invisible under an ordinary light microscope and would appear as a nondescript film using more powerful atomic microscopes. Flat life would also be undetectable in standard biological assays, which rely on the presence of DNA and proteins. Is it possible that layer life is abundant on Earth today, yet remains overlooked?

There's much to learn about the emergence of self-replicating chemical systems. Whether they first formed as chainlike peptides or films, self-replicating molecular systems appear to be a necessary antecedent to life. Nevertheless, a self-replicating metabolism by itself is not sufficient for life as we know it, and many scientists still argue that genetics came first.

16

The RNA World

*It is generally believed that there was a time in the early
history of life on Earth when RNA served as both the genetic
material and the agent of catalytic function.*

Gerald Joyce, 1991

In spite of the elaborate detail of Wächtershäuser's Iron–Sulfur
World, most origin experts dismiss the idea of a purely metabolic
life-form in favor of a genetics-first scenario. In order to reproduce, even the simplest known cell must pass volumes of information
from one generation to the next, and the only known way to store and
copy that much information is with a genetic molecule similar to DNA
or RNA.

No one has thought more deeply about genetics and the origins of
life than Leslie Orgel at the Salk Institute for Biological Studies in San
Diego. His classic 1968 paper, "Evolution of the genetic apparatus," has
guided generations of researchers, and he continues to exert a tremendous influence on origin theory and experiment. Orgel states that the
central dilemma in understanding a genetic origin of life is the identification of a stable, self-replicating genetic molecule—a polymer that
simultaneously carries the information to make copies of itself and
also catalyzes that replication. Accordingly, he catalogs four broad approaches to the problem of jump-starting such a genetic organism.

One possibility is the emergence of a self-replicating peptide of the
kind made by Reza Ghadiri's group at Scripps, or perhaps a protenoid
as championed by Sidney Fox. The idea that proteins emerged first and
then "invented" DNA holds some appeal, because amino acids, the constituents of proteins, are thought to have been available in the prebiotic environment. The problem is that the random prebiotic assembly

of amino acids would have been a messy business, as Fox's critics were quick to point out. Cells have learned how to form neat, chainlike polymers—the proteins essential to life. But left to their own devices, amino acids link together in irregular, undisciplined clusters—hardly the stuff of genetics.

The second of Orgel's possibilities, the simultaneous evolution of proteins and DNA, seems even less likely, because it requires the emergence of not one but two improbable macromolecules.

Graham Cairns-Smith's Clay World scenario provides an intriguing third option, with genetic-like sequences of elements replicating and acting as templates for organic assembly. So far, however, the Clay World scenario is totally unsupported by experimental evidence.

The fourth and favored genetics-origin model of Orgel and many followers is based on a nucleic-acid molecule such as RNA—a single-stranded polymer that acts both as a carrier of information and as a catalyst that promotes self-replication. Orgel proposed this model in 1968, long before any experimental evidence supported such a notion. "I must confess to a strong, longstanding bias in favor of [this] explanation," he remarked recently. "It is, at the very least, the model that can be studied most easily in the laboratory."

How to choose? When evaluating various origin-of-life models, scientists aren't restricted to chemical experiments alone. The metabolism-first models of Wächtershäuser, de Duve, and others are equally influenced by top-down studies of molecular phylogeny, which point to deeply embedded, primordial biochemical pathways. The principle of continuity demands an unbroken path from ancient geochemisty to modern biochemistry. Hence, the citric acid cycle that lies at the heart of all modern metabolism becomes a prime target for studies of protometabolism.

In like fashion, top-down studies of molecular genetics have zeroed in on RNA as the essential core molecule of ancient genetics.

THE RNA WORLD

Few events have electrified the origin-of-life community as much as the early 1980s discovery of RNA ribozymes—strands of RNA that not only carry genetic information, but also act as catalysts. Sidney Altman of Yale and Thomas Cech of the University of Colorado independently demonstrated that a particular segment of RNA can accelerate key biochemical reactions. This startling finding, which won Altman and Cech the Nobel Prize in 1989, inspired a new vision of life's origin.

Modern life relies on two complexly interrelated molecules: DNA, which carries information, and proteins, which perform chemical functions. This interdependence leads to a kind of chicken-and-egg dilemma: Proteins make and maintain DNA, but DNA carries the instructions to make proteins. Which came first? RNA, it turns out, has the potential to do both jobs

The RNA World theory quickly emerged following the discovery of ribozymes. It champions the central role of genetic material in the dual tasks of catalyst and information transfer. Over the years, "RNA World" has come to mean different things to different people, but three precepts are common to all versions of the theory: (1) Once upon a time, RNA rather than DNA stored genetic information; (2) ancient RNA replication followed the same rules as modern DNA replication by matching pairs of bases: A-U (the pyrimidine uracil, whose DNA equivalent is thymine) and C-G; and (3) ancient RNA played the same catalytic roles as modern protein enzymes. In this scenario, the first life-form was simply a self-replicating strand of RNA, perhaps enclosed in a protective lipid membrane. According to most versions of this hypothesis, modern metabolism emerged later, as a means to make RNA replication more efficient.

Two factors may have contributed to the speed with which the RNA World idea caught on. For one thing, a generation after the Miller–Urey experiment, there were still few solid clues about how to make the transition from the prebiotic soup to cellular life. The origin-of-life community was poised to try something new, and RNA provided a compelling original angle, rich in experimental possibilities. In addition, evidence of the dual role of RNA, as both catalyst and carrier, proved seductive to the new generation of biologists, who were born and raised in the age of molecular genetics. To many researchers, life and genetics are synonymous, so the RNA World idea resonates deeply.

The more that biologists learn about RNA, the more remarkably versatile it seems. One big surprise came from the study of ribosomes, lumpy cellular structures that help to assemble proteins. Ribosomes consist of a complex intergrowth of proteins and several RNA strands. Many biologists assumed that the proteins play their usual active role as the enzymes that do the actual assembly work, while RNA merely holds the ribosomes together. However, recent studies prove just the opposite—that RNA mediates the critical step of linking up the protein's constituent amino acids. In essence, RNA does the heavy lifting in protein assembly—a discovery that strongly reinforces RNA's presumed ancient role in biochemistry.

RNA's probable antiquity is underscored by a growing list of other biochemical studies. For example, RNA nucleotides play key structural roles in a variety of essential biological catalysts called coenzymes. These versatile catalysts promote vital reactions at the very heart of the citric acid cycle (the difficult synthesis of citrate from oxaloacetate, for instance). Coenzymes also mediate the manufacture of lipids and other essential biomolecules. And recently, scientists at Yale discovered "riboswitches"—remarkable segments of RNA that change shape when they bind to specific molecules in the cell. These chemical sensors then regulate the cell's chemistry by turning genes on and off.

The inevitable conclusion: RNA is a very ancient molecule that seems to "do it all."

CAVEATS

Today, every origin-of-life meeting features sessions dedicated to RNA World studies. A thousand articles amplify the idea, and hundreds of researchers have pursued variations on the theme. There can be little doubt that the emergence of RNA represents a crucial step in life's origin. However, decades of frustrating chemical experiments have demonstrated that the RNA World could not possibly have emerged fully formed from the primordial soup. There must have been some critical transition stage that bridged the prebiotic milieu and the RNA World.

I am persuaded by those who argue that a self-replicating metabolic system must have emerged first, followed by some form of genetic molecule that was both structurally simpler and chemically more stable than RNA. Only much later did the mechanisms of RNA genetics and ribozymes come into play. Here are some reasons:

1.   Metabolism, which in its earliest stages uses rather simple molecules in the C–O–H (and maybe S) chemical system, seems vastly easier to jump-start than genetics. By contrast, the RNA World scenario relies on exact sequences of chemically complex nucleotides in the C–O–H–N–P system. Accordingly, modern cells synthesize nucleic acids through metabolism, but RNA synthesis is several steps removed

from the core metabolic cycle, the citric acid cycle. This layering of a simple core metabolism surrounded by successively more complex layers of synthesis suggests that metabolism came first and other chemical pathways were added later.

2. Many of the presumed protometabolic molecules are synthesized with relative ease in experiments that mimic prebiotic environments, à la Miller–Urey. RNA nucleotides, by contrast, have never been synthesized from scratch, in spite of decades of focused effort.

3. Even if a prebiotic synthetic pathway to nucleotides could be found, a plausible mechanism to link those individual nucleotides end-to-end into an RNA strand has not been demonstrated. So it's not obvious how catalytic RNA sequences would have formed spontaneously in any prebiotic environment.

Sometimes you have to place your bets and put your cards on the table. I view the RNA World as a critical, but relatively late, transitional stage that occurred when life was well established on Earth—well after the emergence of a stable, evolving metabolic world, and before the modern DNA-protein world. Biologists seem reasonably confident that the last stages of this evolution—the transition from the RNA World to a DNA-protein genetic system—can be understood. Top-down studies of modern life-forms and the genetic code provide abundant clues about that process.

The greater mystery lies in the seemingly intractable gap between primitive metabolism and RNA. Before we can contemplate the RNA World, therefore, we have to address the pre-RNA World. By what chemical process did the first information-bearing system emerge?

17

The Pre-RNA World

I've been waiting all my life for an idea like this.
Simon Nicholas Platts, 2004

What preceded the RNA World? We understand a lot about the possible earliest stages of life's emergence—how to make the prebiotic soup with all sorts of interesting molecules and how to assemble those molecules into a variety of larger useful structures. At the other end of the story, we have a good handle on how strands of RNA might function as evolving, self-replicating systems (as we'll see in the next chapter). But there's that maddening gap between the primordial soup and the RNA World. Stanley Miller sums up the problem: "Identifying the first genetic material will provide the key to understanding the origin of life. RNA is an unlikely candidate."

To be sure, there have been numerous creative attempts to close this gap. Several researchers have approached the problem by proposing simpler types of precursor genetic polymers that might have arisen before RNA. In a *tour de force* research program, the Swiss chemist Albert Eschenmoser explored the stabilities of more than a dozen variants of RNA with modified sugar-phosphate backbones. He systematically replaced the 5-carbon sugar ribose with various other 4-, 5- and 6-carbon sugars and discovered seven new kinds of stable nucleic-acid-like structures. Most significant was the discovery by Eschenmoser and colleagues of a nucleic acid with the 4-carbon sugar threose (the molecule was dubbed TNA). Unlike ribose, which must be synthesized through a rather cumbersome sequence of chemical steps, threose can be assembled directly from a pair of 2-carbon molecules. This difference makes TNA a much more likely molecule than RNA to arise spontaneously from the prebiotic soup.

Other scientists took a different chemical tack. In 1991, the Danish chemist Peter Nielsen and colleagues synthesized a novel genetic molecule—a "peptide nucleic acid" (PNA), which features RNA-like bases bound to a backbone of amino acid molecules. The reliance on readily available amino acids, rather than problematic sugar phosphates for the polymer backbone, appealed to many members of the origins community. The discovery of PNA also underscored the chemical richness of plausible genetic molecules.

These immensely creative efforts expand the repertoire of prebiotic possibilities. They also hold the promise of providing new kinds of synthetic genetic molecules that can interact with modern cells yet not interfere with cellular function—a potential boon to medical research. Nevertheless, no one has managed to achieve a plausible prebiotic synthesis of these alternative nucleotides, much less a viable genetic polymer. The door is wide open for new ideas.

THE PAH WORLD

WARNING: The following section presents an intriguing hypothesis, but one that is highly speculative and as yet untested. Such novel ideas fuel origins research, though most are eventually cast aside—the victims of faulty chemical reasoning or failed experimental predictions. Whatever the outcome, this story epitomizes the exhilarating process of scientific exploration.

In May of 2004, Simon Nicholas ("Nick") Platts was in trouble. After almost five years of fruitless graduate work, and approaching his fortieth birthday, he was about to be deported to his native Australia with no degree and no job. [Plate 6]

Nick's life had defied the traditional arrow-straight science career path that most of us knew: college, graduate school, postdoc, and tenure track. A gifted chemist and enthusiastic educator, he went from a master's degree in chemistry to teaching at Melbourne Grammar School. Only at the advanced age of 35 did he resolve to return to university, get a doctorate, and do some research. Life's chemical origin was one topic that really excited him, so he applied to Rensselaer Polytechnic Institute in Troy, New York, to work with Jim Ferris.

His new RPI colleagues found Nick to be outgoing and supportive, always ready to help others with their research, and eager to teach undergraduate chemistry, but progress on his thesis project suffered. After three years in Troy, he moved to Washington, D.C., and the Carnegie Institution as a NASA-sponsored predoctoral fellow, a position that would allow him to acquire his doctorate from RPI, but do the research in a new setting with fewer distractions. He had two years left to complete a thesis.

Nick was full of ideas. Early in 2003, he came to me with plans for an elegant experiment on the possible influence of Earth's magnetic field on the origin of biological homochirality. Could I provide some lab space?

"OK, but for how long?" I asked. My lab is small and space is at a premium.

"Once it's set up, the experiment should take only a couple of weeks," he assured me, so I offered him the necessary space for a few months.

"No worries!" his Aussie reply.

Nick started with a flurry of activity, marking out a good fraction of my lab bench with masking tape, cordoning off a sink that might otherwise splash onto the lab's delicate apparatus, assembling an elaborate optical table, and filling cabinets with hardware, but his progress soon slowed. Crucial supplies were on back order, funds were needed for a special laser, other difficulties followed. More than a year later, the sequestered lab space, still piled high with equipment, was gathering dust. When pressed about his plans, Nick was vague and evasive. Many of us feared that he would be forced to leave the Geophysical Lab empty-handed at the end of June 2004. A sense of shared responsibility weighed heavily on George Cody, Marilyn Fogel, and me.

EUREKA!

Nick Platts' life changed dramatically for the better on Tuesday, May 25, 2004. That's when the central idea for his PAH World scenario crystallized. Max Bernstein, a colleague of Lou Allamandola at NASA Ames, had planted the seeds for the concept at an RPI seminar in November 2001. Bernstein's talk had emphasized the occurrence of polycyclic aromatic hydrocarbons in deep space, and it got Nick thinking about PAH chemistry. A September 2003 conference in Trieste in celebration of the 50th anniversary of Stanley Miller's breakthrough experiment inspired more progress. "On the return flight," Nick remembers, "I scribbled the idea on the back of a United Airlines ticket."

The germ of an idea gradually developed "on the mental back burner," but May 25th was the Eureka moment. "It was 5:51 p.m.," Nick recalls. "That's when I did the drawings. That's when I realized how big this was." Contrary to popular myth, there aren't many such moments in science.

Nick's new idea follows in the tradition of other models of pre-RNA genetic polymers, but with an original chemical twist. PAHs would have been abundant among the plethora of prebiotic molecules. (We met them back in Chapters 3 and 5 as significant components of carbonaceous meteorites and, by extension, the cosmic debris that formed our planet.) Each of these flat sturdy molecules consists of fused 6-member rings of carbon atoms with hydrogen atoms around the edge. PAHs are relatively insoluble in ocean water, but under the influence of solar ultraviolet radiation they can be chemically modified, or "functionalized"; hydrogen atoms can be stripped off, and new molecular fragments can take their place. If the PAH edges become decorated with OH, for example, then their solubility in ocean water increases significantly.

By any calculation, PAHs and their functionalized variants were a significant repository for organic carbon on the primitive Earth. I suspect that some origins chemists thought it a shame to lock up so many potentially useful carbon atoms in the relatively inert, biologically useless PAHs.

But what if PAHs played a key role in the ancient ocean? What Nick realized was that a functionalized PAH with OH around the edges is an amphiphile—a molecule that both loves and hates water, similar to Dave Deamer's lipids (see Chapter 11). The flat surfaces of the carbon rings repel water (they're hydrophobic), but the OH-bearing edges of the PAHs attract water (they're hydrophilic). So what happens when lots of functionalized PAHs are placed in seawater? How might the hydrophobic parts stay away from water, while the hydrophilic parts remain in contact with the wet surroundings? The simple answer is to assemble a pile of PAHs like a stack of plates in the cupboard. The flat hydrophobic parts shield each other from water, while the edges form the water-bathed outside of the stack. The PAHs self-organize.

Platts predicted that, once stacked, the system would preferentially bind small flat molecules, like the DNA bases, to the PAH edges and thus concentrate them from the surrounding prebiotic soup. These baselike molecules would attach and break free in a constant game of molecular musical chairs. Gradually, however, a selection process would take place. Because the PAHs are loosely stacked, adjacent flat PAH molecules would slide back and forth and rotate next to each other, like your hands do when you rub them together on a cold day. This mechanical action would tend to break off any edge-bound molecule that isn't, itself, as flat as a PAH. Consequently, the small, flat, ring-shaped base molecules (the A, C, G, and U of RNA, for example) would bind preferentially around the edges of the stack. What's more, these RNA bases are also amphiphilic, so they might have a tendency to line up more or less on top of each other, forming a kind of ministack. Remarkably, the spacing between the PAH layers (and hence the vertical separation between bases) is 0.34 billionths of a meter—exactly the same spacing as found between the bases of DNA and RNA.

The result of all this self-organization, according to Nick's PAH World hypothesis, would be a stack of PAHs decorated along the side by an array of small flat molecules, including bases—an arrangement that looks for all the world like the information-rich genetic sequence of DNA or RNA.

Once bases are effectively stacked, Platts suspects, other small molecules would start to form a backbone linking the bases together into a true information rich molecule, though those key chemical steps remain fuzzy. Then a change in the ocean water's acidity, as might be experienced by moving from a deep hydrothermal environment to a more shallow zone, might allow the sequence of bases to break free as a true pre-RNA genetic molecule that could fold back on itself to match up base pairs. Eventually, complex assemblies of these polymers might act as catalysts for self-replication and growth.

Nick's proposal lacked any experimental support. Nevertheless, the PAH World hypothesis seemed to provide a geochemically plausible, self-consistent, conceptually simple, and seamless chemical path from the dilute soup to an RNA-like genetic sequence.

SHARING

Nick Platts' model crystallized in a flash, but was it reasonable? Did it make sense? He decided the best strategy was to bounce the idea off other people.

A

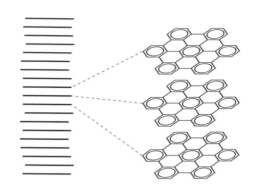

B

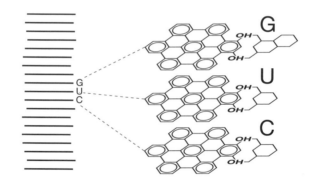

C

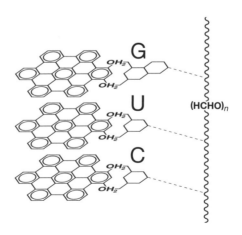

D

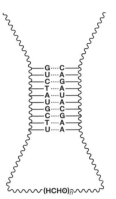

E

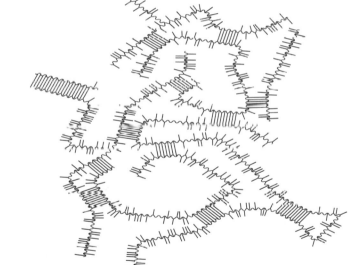

(A) Nick Platts' PAH World hypothesis rests on the ability of polycyclic molecules to self-organize into stacks. (B) Once stacked, the PAHs would tend to attract small flat molecules (notably the bases of DNA and RNA) to the edges. (C) A molecular backbone forms to link the bases into a long chain. (D) The RNA-like chain of bases separates from the PAHs and folds into a molecule that carries information. (E) Complex assemblages of these chains have the potential to catalyze reactions. These drawings are adapted from Nick Platts' unpublished manuscript.

I must have been about tenth on his list. Nick appeared at my office door on the afternoon of Thursday, May 27th. "You got a few minutes?" he asked, taking a seat. I nodded, hoping for a progress report on the stalled thesis experiments. "I've found something extraordinary," he began. "I think I've discovered how life began."

And so he described his hypothesis—the self-assembly of functionalized PAHs, the selection of flat molecules along the edges, the fortuitous spacing of the bases. He sat on the edge of his chair, leaning toward me and gesticulating as he spoke. Nick was clearly exhausted from almost two days without sleep; his voice took on a manic intensity. "This is a once-in-a-lifetime moment! I've never been part of anything this big!"

I was a bit taken aback by what sounded like a wildly speculative idea. It seemed at first like another distraction, just weeks before his scheduled doctoral defense.

"I told Dave Deamer and he loves the idea." Nick rattled off the names of half a dozen other origin-of-life experts he'd already contacted. "No one can think of an objection." Then, paradoxically, "We've got to keep this secret. Someone else will be sure to steal the idea, so don't tell anyone."

"Can you propose any experimental tests?" I asked him. A safe, neutral question, while I considered how else to respond.

He deflected the question. "There's lots we can do, but we have to get this out fast. I've been drafting a manuscript. I'd like you to be a coauthor. Where do you think we should submit it?"

A manuscript? Coauthorship? Nick had just raised the stakes. I was uncomfortable with being an author unless I could contribute something original to the paper, but I was happy to discuss publication strategy. I thought the ideal forum for a short, high-impact outline would be a 700-word "Brief Communication" in *Nature* or a similarly concise "Brevia" in *Science. Proceedings of the National Academy of Sciences* was another option, but *PNAS* articles are generally longer, and Nick had no data yet with which to flesh out his hypothesis. He agreed to adopt the short *Nature* format, while I promised to read his paper and comment quickly.

The next day, Nick e-mailed me a 700-word draft for "Edge-derivatised and stacked polycyclic aromatic hydrocarbons (PAHs) as essential scaffolding at the origin of life." I could tell from the title that the paper was going to need some work. Even so, as I read the text I warmed to the elegant theory.

The hallmark of any useful scientific hypothesis is that it makes unambiguous predictions. Nick's hypothesis made testable predictions by the bundle. First, functionalized PAHs must self-assemble into stacks. The stacked PAHs, furthermore, must be of similar size and shape. PAH edges must attract a variety of molecules, but there must be a preferential selection for flat, baselike molecules. And the bases must also be aligned vertically. If only we could confirm at least one of these predictions.

George Cody provided the chemical evidence that made me a believer. Coal, George's specialty, is loaded with PAHs. It turns out that there's already an extensive scientific literature on the ability of functionalized PAHs to self-organize into stacks—a process known as discotic organization. A quarter-century of publications had already elaborated on Nick Platts' prediction. Neither Nick nor I had ever heard the word "discotic" until Cody mentioned it, drawing on seminars he'd heard on the subject while working at Exxon. I returned to Nick's manuscript with a red pen and renewed intensity. My principal contribution was to come up with a catchier title, "The PAH World."

By Tuesday, June 1st, Nick had received comments from more than a dozen scientists from around the world and the draft manuscript was taking final form. He called a meeting of Carnegie coauthors—nine of us in all—at the Lab's library. We sat around a massive mission oak table and worked through the paper one last time. Should we talk about the common mineral graphite, which also has flat carbon rings? Had we included the most appropriate references? Should we propose specific experiments? Our biggest concern was Nick's figure, which needed to make the essentials of the model as clear as possible, and he agreed to redo it. He submitted the PAH World paper to *Nature* on Thursday, June 3rd, just nine days after his epiphany.

OPINIONS

Nick had no time to relax. Whether or not *Nature* accepted the paper, he wanted us to stake a claim. We began e-mailing the manuscript, designated "in review," to a dozen astrobiologists and origins experts to request their comments. We focused on leading RNA World proponents—Jerry Joyce, Leslie Orgel, Jack Szostak—figuring they would have the most to gain from the novel idea.

Responses came quickly and with varying degrees of enthusiasm. A few scientists who were good friends of the Lab were warmly congratulatory, but most respondents remained cautious, and almost everyone cited the need for experimental evidence. Jack Szostak responded in less than an hour with some skepticism, adding "I think it's worth pursuing experimentally—it would certainly be cool if an effect could be demonstrated." The next day Leslie Orgel chimed in, "An experimental demonstration of your scheme might be interesting, but I wouldn't advise publishing without showing that it works well."

"I thought it was interesting and certainly appealing," echoed Jerry Joyce. "However, it would help tremendously if there were some experimental support."

Andy Knoll joined the chorus: "For now it is fascinating speculation, but the ideas seem amenable to what you guys do best—careful experimentation."

Evidently the editors at *Nature* agreed. We received the form rejection letter a week later. "Thank you for submitting your contribution, . . . which we are unable to consider for publication." Boilerplate didn't make their decision any easier to swallow. "Because of severe space constraints . . . we are unable to return individual explanations to authors. . . ."

EXPERIMENTS

Experimental evidence seemed the obvious key to securing support for the PAH World hypothesis. Unfortunately, experiments take time, and that was one commodity of which Nick Platts had precious little at this stage in his graduate work.

In such circumstances, it's best to think boldly. Nick envisioned a single sequence of experiments that might validate every facet of his theory. His plan: First obtain a sample of a modest-sized PAH, one with a dozen or so interlocking rings. He settled on the elegant, symmetrical hexabenzocoronene (HBC), with its starlike pattern of 13 rings. Put the HBC into a flask with some water, functionalize the molecules by irradiation with an ultraviolet lamp, then measure the system for discotic organization. That much we were pretty confident would work, based on a survey of the discotics literature. Then add base molecules, such as adenine and guanine (the A and G of the RNA alphabet to the system and see if the PAHs bond to them. Finally—and we all knew that this would be a long shot—add a molecule that might serve as a backbone, such as formaldehyde, and see if the bases would link together into a long sequence reminiscent of a genetic polymer.

The experimental protocols looked good on paper, but there were problems from the start. Pure PAHs (as opposed to random sooty mixtures) turn out to be almost impossible to find commercially, and when you do, they're impossibly expensive. Nick located one European supplier who would sell us tiny amounts of HBC for a thousand dollars—the equivalent of more than a million dollars for ten grams! That clearly wasn't going to work.

Nick contacted other labs in Europe and Japan, hoping for a complimentary supply, but tracking one down might take weeks or months. Meanwhile, he manufactured some PAHs of his own by burning acetylene in air and catching the sooty residues in one of my more expensive glass beakers. Gradually, the soot, which is a mixture of hundreds of different PAHs ranging in size from a few rings to hundreds, piled up.

I offered to try another angle: purifying PAHs from soot by TLC—the thin-layer chromatography technique I had learned in Dave Deamer's lab (see Chapter 11). Nick agreed, and we ordered a supply of trichlorobenzene, one of the solvents of choice when working with PAHs.

Nick had other concerns besides experiments. He was still determined to publish his idea, so he prepared a new version of the short paper and submitted it to *Science* on Thursday, July 15th.

Lots of people at Carnegie had been aware of the PAH World buzz and wanted to hear from Nick first hand. He presented an informal talk to the Carnegie campus on Friday afternoon July 2nd and again for a NASA Astrobiology Institute video seminar on Monday, July 12th. Each talk provided an opportunity to hone his arguments and to field a new round of questions and comments from origins experts across the spectrum.

His thesis defense also loomed large. Scrambling to write up an expanded description of his hypothesis and assemble whatever support he could from the published literature, Nick cobbled together a doctoral dissertation and headed back up to Troy for a July 20th defense. We all wondered whether he could pull it off. As it turned out, his thesis committee, chaired by chemistry professor Gerry Korenowski, was more than a little impressed by Nick's elegant vision. In five years of graduate work he had demonstrated his depth of chemical understanding and intuition. The PAH World hypothesis established beyond doubt that he had earned the right to be called "Doctor Platts." They granted Nick an extra week to polish up his prose, and on July 27th he successfully defended what has to be one of the most unusual chemistry theses in RPI's history.

MOVING ON

Having his doctorate in hand was a tremendous relief, but Nick's problems weren't over. His visa was about to run out unless he could find a science job in the United States. Could I help?

Nick's inspiration centered on the behavior of self-organizing amphiphilic molecules. The logical person to call was Dave Deamer, who received some support from our NASA Astrobiology Institute grant to study molecular self-organization. Dave had also published some intriguing speculations about the energetic role of functionalized PAHs, which might have gathered light to power a sort of primitive photosynthesis. I sent Dave a long e-mail describing Nick's unusual circumstances and his special gifts. Dave, always kind and gracious, responded almost immediately. Yes, he thought he would have a research slot opening up in the fall. When could Nick start?

Nick is in Santa Cruz now, attempting the experiments that may support or refute his ingenious model. It's too soon to tell if the PAH World hypothesis will pan out, but whatever the outcome of Nick Platts' work, we'll know more than we did before. PAHs, or PNA, or some other information-rich, self-replicating genetic system must have emerged on the path to the RNA World. That first genetic molecule must have been much more resilient than RNA, though it was undoubtedly significantly less efficient as a replicator. At some point, as self-replicating genetic molecules became more competitive, RNA took over. The era of Darwinian evolution had begun.

18

The Emergence of Competition

It is evident that once a self-replicating, mutating molecular aggregate arose, Darwinian natural selection became possible and the origin of life can be dated from this event. Unfortunately, it is this event about which we know the least.

Carl Sagan, 1961

Life's most poignant hallmark is inescapable, inexorable change. Each of us is born, grows old, and dies. Each species arises from prior species, fills its own niche for a time, and eventually becomes extinct. In such a sweeping, undirected evolutionary drama, the human species might be seen as but one small, insignificant player. Little wonder that Darwin's theory of evolution by natural selection has met with so much hostility.

The process of evolution by natural selection rests on two incontrovertible facts. First, every population is genetically diverse, possessing a range of traits. Second, many more individuals are born than can hope to survive. Consequently, over time those individuals with more advantageous traits are more likely to survive and pass on their genetic characteristics to the next generation. This selection process drives evolution.

Even today, almost a century and a half after the publication of *The Origin of Species*, a vocal minority of Americans views the theory of evolution as a dangerous and subversive doctrine that substitutes materialism for faith. Efforts to expunge Darwin from textbooks and to augment curricula with thinly veiled religious beliefs in the guise of "scientific creationism" or "intelligent design" continue unabated in many states.

But natural selection is not a sinister development in life's emergence. On the contrary, evolution is the natural and necessary sorting-out process that led to the origin of life.

MOLECULAR SELECTION AND EVOLUTION

Imagine yourself back to the primitive Earth more than 4 billion years ago, to a time before life had emerged. Oceans and shorelines must have held a bewildering, chaotic diversity of organic molecules from many sources. Somehow that confused chemical mess had to be sorted out. Two connected processes—molecular selection and molecular evolution—winnowed and modified the prebiotic mélange.

Molecular selection, by which a few molecules earned starring roles in life's origin, proceeded on many fronts. Some molecules were inherently unstable or unusually reactive and so they quickly disappeared from the scene. Other molecules proved to be soluble in the oceans and so were removed from contention. Still other molecular species sequestered themselves by bonding strongly to surfaces of chemically unhelpful minerals or clumped together into gooey masses of little use for the emergence of life.

Earth's many cycles amplified these emergent selection processes. Tidal pool cycles of wetting and evaporation concentrated molecules, while the Sun's ultraviolet radiation fragmented the least stable molecular species. Pulses of hydrothermal seawater delivered new supplies of chemicals to deep-ocean vents, where differential adsorption and detachment of molecules on reactive mineral surfaces concentrated a select subset of molecular species. Day and night, hot and cold, sun and rain, high tide and low—these and other periodic phenomena refined the chemical mix.

In every geochemical environment, each kind of organic molecule had its reliable sources and its inevitable sinks. For a time, perhaps for hundreds of millions of years, a kind of molecular equilibrium persisted, as the new supply of each molecular species was balanced by its loss. Such an equilibrium assemblage features nonstop reactions among molecules, to be sure, but the system does not necessarily evolve.

At some point, by processes as yet poorly understood, a self-replicating cycle of molecules emerged, thereby changing the character of molecular selection. Even a relatively unstable collection of molecules could persist in significant concentrations if it made copies of itself. Molecules within that first self-replicating collection would have thrived at the expenses of other molecular species.

Theorists imagine a more interesting possibility—an ancient environment in which two or more self-replicating cycles of molecules competed for atoms and energy. Such competition inevitably arose as new molecular species offered alternative chemical pathways, or perhaps as changes in environment triggered slight variations in a cycle. Dueling molecular networks would have vied for resources, mimicking life's unceasing struggle for survival. In such a competitive environment, increasingly efficient cycles emerged and flourished at the expense of less efficient variants, slowly shifting the molecular balance. Molecular evolution had begun on Earth.

The dynamic, competitive tussle of molecular evolution differs fundamentally from the more passive process of molecular selection. Competition among self-replicating cycles drives evolutionary change, fostering efficiency and introducing novelty. That's why many origin experts draw the arbitrary line that separates living from nonliving systems at the emergence of a self-replicating chemical cycle that began to evolve by this powerful process of natural selection.

Replication of a cycle, in and of itself, is not enough to claim life. The cycle must also possess a sufficient degree of variability so that when it competes for resources, more efficient variants win out. Over time the system changes, becoming more adept at gathering atoms and energy. Under these conditions, the emergence of increasing molecular complexity is inevitable, as new chemical pathways overlay the old. So it is that life has continued to evolve over the past 4 billion years of Earth history.

EVOLUTION IN THE LAB

A theory, no matter how plausible, requires experimental testing, but how can one test molecular evolution in a laboratory? That challenge seems almost beyond imagining, yet that is exactly what a few teams of biochemists have done.

Laboratory studies of molecular evolution began in the 1960s at the University of Illinois, where Professor Sol Spiegelman and his colleagues investigated a tiny virus called Qβ (pronounced "que-beta") that attacks the common bacterium E. coli. Qβ is nothing more than a protein shell surrounding a small loop of RNA. Its entire existence is devoted to infecting *E. coli* cells and inducing those cells to make more copies of the Qβ virus. To do that, one of Qβ's proteins, Qβ replicase, has to make copies of Qβ RNA.

This close relationship between Qβ replicase and Qβ RNA suggests a promising experiment. If you put some Qβ replicase and a Qβ RNA strand into a test tube along with lots of small RNA building blocks, you'll wind up with countless new Qβ RNA strands. One caveat: Qβ replicase makes lots of mistakes, so each new RNA strand is likely to vary slightly from the original.

Here's what Spiegelman and co-workers did to make their molecules evolve. They allowed the chemical mix to make copies for just 20 minutes. During that short period, some RNA sequences bound strongly to Qβ replicase and duplicated rapidly, while others interacted poorly with Qβ replicase and were inefficiently copied. After 20 minutes, when the solution had become enriched in RNA strands that easily replicate, they transferred a small amount of that liquid into a new beaker with fresh Qβ replicase and RNA building blocks. Then they ran the experiment for another 20 minutes and transferred a small amount of the second batch of RNA copies into a third beaker with fresh Qβ replicase.

Seventy-four times they repeated the process, gradually reducing the time allowed for the reaction to proceed. Each step preferentially selected those RNA strands that were copied most efficiently. By the end of the experiment, the length of the most efficient RNA strands had been shortened to a sixth of the original size, and these evolved RNA sequences were being copied 15 times faster than before. Similar experiments generated strands of RNA that were unusually resistant to high temperature or to the effects of damaging chemicals. Under such severe selection pressure (time, temperature, or some other stress), the Qβ RNA evolved.

SELF-REPLICATING RNA

"Spiegelman monsters," as these systems came to be known, are not alive. Only by the most overt manipulation (cycling through a succession of beakers, for example) does the Qβ RNA evolve. But the repetitive selection process developed by Spiegelman's group proved that molecular systems under competitive pressure can be induced to evolve—a result that suggested to many researchers a brave new world of artificially evolved molecules designed to accomplish specific chemical tasks.

Harvard Medical School biologist Jack Szostak has set his sights even higher. Indeed, Szostak's overriding ambition is nothing less than to design an evolving life-form in his lab. "Our ultimate goal is to create life," he admits.

Jack impresses you as someone who is supremely happy in his profession. Unassuming and quiet by nature, he often wears a half-smile, as if he were amused by a private joke. Despite his 50-odd years, his slender build and the bright eyes behind round-framed glasses make him look like a scientific Harry Potter—and he is in fact a bit of a wizard.

For Jack, RNA is the key. He began his scientific career studying yeast chromosomes—an important mainstream task in molecular biology that sheds light on how DNA operates in humans. Chromosomes are elongated structures that divide and separate during cell division; they carry all of yeast's DNA. Szostak and his student, Andrew Murray, synthesized one from scratch. "Even then," he recalls, "my guiding principle was that the best way to show that you really understand how something works is to try to build it from pieces and then see if it works as expected." [Plate 8]

The study of yeast DNA was a crowded field, and Szostak itched to try his hand at something different. The discovery in the early 1980s of ribozymes—RNA that behaves as a catalyst—was too seductive to pass up, and he soon changed research directions.

In their first series of ribozyme experiments, Szostak and a group of his students focused on engineering chains of RNA that can copy other RNA molecules. Such an "RNA replicase" represents a first key step in designing a self-replicating chemical system. Success came in 1989, when he and Jennifer Doudna (now a professor at Berkeley) made an RNA molecule that copies short RNA sequences, albeit rather sloppily. Other high-profile papers soon followed, as their ability to manipulate RNA improved. These were spectacular advances, but still a long, long way from a reliable self-replicating RNA strand. They'd have to do better, but how? Jack decided to let molecular evolution work for him.

Szostak's objective was to evolve an RNA molecule that would attach strongly and selectively to a "target molecule" of distinctive shape. His team tackled RNA evolution by first generating a solution with more than 10 trillion random RNA sequences, each about 120 "letters" long. They poured this RNA-rich solution into a beaker whose glass sides had been coated with the target molecule. After sitting for a few minutes, the vast majority of RNA chains had done nothing, but a few RNA strands, by chance, were able to grab onto the target molecule and thus remained firmly attached to the beaker. When they flushed the RNA solution out of the beaker, those relatively few RNA sequences that bonded to the target molecule remained behind.

At first, only the tiniest fraction of the random RNA sequences attached to the target, but Szostak's team collected those precocious strands and made trillions of approximate copies—similar sequences but with lots of random mutations thrown in. Then they repeated the experiment and picked out a second generation of RNA sequences that did the job better than those from the first. Again and again, they cycled the RNA, each time copying the best sequences and thus improving the speed and accuracy with which the RNA latched onto the glass-bound target molecules. After several dozen cycles, the surviving RNA strands had evolved to the point where they were perfectly adapted to the assigned task. This elegant evolutionary process has now been extended to the design of numerous new RNA sequences with a wide range of specialized functions, from locking onto viruses to splicing DNA.

With their new procedures, Jack Szostak and his colleagues have turned their attention to self-replicating RNA. In 2001, David Bartel, a former Szostak graduate student now at MIT and the Whitehead Institute, managed to produce an impressive RNA sequence that can grab onto a shorter piece of RNA up to 14 letters long and copy it. Optimistic researchers are convinced that it's only a matter of time before self-replicating RNA strands more than a hundred letters long will be commonplace. If so, then synthetic life may not be far behind.

SYNTHETIC LIFE

Self-replicating RNA is not alive, but it's getting closer. It's a big molecule that carries genetic information, catalyzes its own reproduction, and mutates and evolves to boot. But no plausible geochemical environment could feed such an unbound molecule, nor would it have survived long under most natural chemical conditions.

One key to survival is protection, and that's where a lipid membrane comes in. Building on David Deamer's discoveries, Szostak protégés Martin Hanczyc and Shelley Fujikawa have experimented with ways that membranes might have encapsulated RNA strands. Among their findings: Lipid vesicles that are squeezed through tiny pores stretch out, divide, and start to grow larger—a process that mimics cell division. What's more, the process is greatly accelerated by the addition of fine-grained clay minerals, some of which end up inside the vesicles. Recall that Jim Ferris at RPI demonstrated that clays can attract and help to assemble RNA strands. So, in the spirit of Pier Luigi Luisi's Lipid World, it might be possible to make cell-like structures that spontaneously incorporate RNA-bound clay particles.

This behavior of lipids and RNA has led Szostak and his students to propose a remarkable scenario for the first life-form to evolve by natural selection. Imagine a lipid vesicle that contains self-replicating RNA. Previous authorities have suggested that RNA must have played many roles in such a protocell (roles that DNA and proteins play today)—manufacturing new membrane molecules, controlling cell shape and size, copying itself, and more.

Szostak's team realized that RNA could drive cell growth by the much simpler process of internal pressure. RNA pushes out on the membrane, which in turn presses on neighboring protocells. They speculated that this contact would promote the transfer of lipids from cells with less internal pressure (hence, less RNA) to those with more. Thus the competition for space would lead to a natural-selection process. Protocells with more RNA would be more successful.

To test their ideas, they first prepared one set of vesicles filled with a solution of the sugar sucrose and another set filled with pure water. When mixed together and confined, the sucrose-ladened vesicles grew larger by drawing lipids from their sugarless neighbors. Repeating the experiment with RNA strands yielded the same results. Vesicles swollen with RNA grew, as the adjacent empty vesicles shrank.

Previous workers had assumed that RNA would have to learn how to accomplish several tasks—lipid synthesis, self-replication, metabolic functions, and more—before a protocell could evolve by natural selection. Szostak's latest results suggest a much simpler scenario, in which the only essential task for protocell competition is RNA self-replication. "If we can get self-replicating RNAs," Szostak suggests, "then we can put them into these simple membrane compartments and hope to actually see this competitive process of growth." The more

RNA a vesicle captures and copies, the more successfully it will compete with its neighbors—a sort of molecular the-rich-get-richer scheme. In such a world, the most efficient RNA replicators would enjoy a tremendous advantage. With the emergence of competition, Darwinian evolution could take center stage.

Every year, Jack Szostak moves closer to his goal of creating a self-replicating, encapsulated, evolving chemical system in the lab. If and when he or his successors accomplish this, it will be a historic achievement. Synthetic life will also trigger a new flood of ethical questions about the potential dangers of scientific research, as well as philosophical questions about the meaning of life. But will synthetic life tell us how life emerged on Earth?

A synthetic RNA organism will certainly give credibility to the RNA World hypothesis that a strand of RNA (or some precursor genetic molecule) formed the basis of the first evolving, self-replicating chemical system. But laboratory-created life will not have emerged spontaneously from chemical reactions among the simple molecular building blocks of the prebiotic Earth. Researchers still stack the deck by supplying a steady source of RNA nucleotides and vesicle-forming lipids. And so, for the time being, deep mysteries remain.

19

Three Scenarios for the Origin of Life

Anyone who tells you that he or she knows how life started on
the sere Earth some 3.45 billion years ago is a fool or a knave.
Stuart Kauffman, *At Home in the Universe*, 1995

What do we know for certain? Scientists have learned that abundant organic molecules must have been synthesized, and must have accumulated, in a host of prebiotic environments. They have also demonstrated many processes by which biomolecular systems—including lipid membranes and genetic polymers—might have formed on mineral surfaces. As molecular complexity increased, it seems plausible that simple metabolic cycles of self-replicating molecules emerged, as did self-replicating genetic molecules.

So we've learned a lot, but what we know about the origin of life is dwarfed by what we don't know. It's as if we were trying to assemble a giant jigsaw puzzle. A few pieces clump together here and there, but most of the pieces are missing and we don't even have the box to see what the complete picture is supposed to look like.

The greatest mystery of life's origin lies in the unknown transition from a more-or-less static geochemical world with lots of interesting organic molecules to an evolving biochemical world in which collections of molecules self-replicate, compete, and thereby evolve. How that transition occurred seems to boil down to a choice among three possible scenarios.

1. *Life began with metabolism, and genetic molecules were incorporated later:* Following Günter Wächtershäuser's hypothesis, life began autotrophically. Life's first building blocks were the simplest of molecules, while minerals provided chemical energy. In this scenario, a

self-replicating chemical cycle akin to the reverse citric acid cycle became established on a mineral surface (perhaps coated with a protective lipid layer). All subsequent chemical complexities, including genetic mechanisms and encapsulation into a cell-like structure, emerged through natural selection, as variants of the cycle competed for resources and the system became more efficient and more complex. In this version, life first emerged as an evolving chemical coating on rocks.

The true test of this origin scenario rests on chemical synthesis experiments. Starting with simple molecules and common minerals subjected to plausible prebiotic conditions, researchers must discover a way to jump-start a self-replicating cycle of molecules that mimics the core citric acid metabolism of modern life-forms. We're close, and several of the essential steps have been accomplished. A key missing experiment is the synthesis of 4-carbon oxaloacetate from 3-carbon pyruvate, perhaps using sulfur analogs in an environment rich in hydrogen sulfide. If that step can be demonstrated, and a self-sustaining cycle of reactions maintained, then metabolism-first will be the model to beat.

2. *Life began with self-replicating genetic molecules, and metabolism was incorporated later:* According to the RNA World hypothesis, life began heterotrophically and relied on an abundance of molecules already present in the environment. Organic molecules in the prebiotic soup, perhaps aided by clays or PAHs or some other template, self-organized into information-rich polymers. Eventually, one of these polymers (possibly surrounded by a lipid membrane) acquired the ability to self-replicate. All subsequent chemical complexities, including metabolic cycles, arose through natural selection, as variants of the genetic polymer became more efficient at self-replication. In this version, life first emerged as an evolving polymer with a functional genetic sequence.

This scenario, with its appealing reliance on the multiple ancient roles of RNA, lacks only the crucial support of an experiment that demonstrates the plausible prebiotic synthesis of a genetic polymer—RNA or its precursor. If an experiment successfully demonstrates the facile synthesis of such a polymer from prebiotic building blocks, then the biggest gap in our understanding of life's origin will have been filled. In that case, many experts will conclude that the problem of life's chemical origin has been solved.

3. *Life began as a cooperative chemical phenomenon arising between metabolism and genetics:* A third scenario rests on the possibility that neither protometabolic cycles (which lack the means of faithful self-replication) nor protogenetic molecules (which are not very stable and lack a reliable source of chemical energy) could have progressed far by themselves. If, however, a crudely self-replicating genetic molecule became attached to a crudely functioning surface-bound metabolic coating, then a kind of cooperative chemistry might have kicked in. The genetic molecule might have used chemical energy produced by metabolites to make copies of itself, while protecting itself by binding to the surface. Any subsequent variations of the genetic molecules that fortuitously offered protection for themselves or for the metabolites, or improved the chemical efficiency of the system, would have been preserved preferentially. Gradually, both the genetic and metabolic components would have become more efficient and more interdependent.

Such a "dual origins" model might at first seem to introduce a needless complication (not to mention sounding like a wishy-washy compromise). Nevertheless, exactly this kind of symbiotic coupling of metabolism and genetics is now thought to have occurred early in the history of cellular life. Crucial features of our own cells suggest an ancient cooperative merging of early, more primitive cells. If experiments establish easy synthetic pathways to both a simple metabolic cycle *and* to an RNA-like genetic polymer, then such a symbiosis may provide the most attractive origin scenario of all.

We don't yet know the answer, but we're poised to find out. Each day, new experiments expose more of the truth and winnow the possibilities. Each day, we get closer to understanding. And whatever the correct scenario, of one thing we can be sure: Ultimately, competition began to drive the emergence of ever more elaborate chemical cycles by the process of natural selection. Inexorably, life emerged, never to relinquish its foothold on Earth.

Epilogue

The Journey Ahead

Once to every man and nation comes the moment to decide,
In the strife of truth with falsehood, for the good or evil side;
Some great cause, some new decision, offering each the bloom or blight,
And the choice goes by forever 'twixt that darkness and that light.
James Russell Lowell, 1845

The theory of emergence points to a gradual, inexorable evolution of the cosmos, from atoms to galaxies to planets to life. Each emergent step arises from the interactions of numerous agents and yields an outcome much greater than the sum of its parts. Each emergent step increases the degree of order and complexity, and each step follows logically, sequentially from its predecessor.

We recognize this majestic progression only in hindsight. Emergent phenomena remain elusive—exceedingly difficult to predict from observations of earlier stages. Given hydrogen atoms, a tremendous conceptual leap is required to predict the brilliance of stars or the variety of planets. Given planets, no theoretician alive could predict the emergence of cellular life in all its diversity—nor, given cellular life, could anyone foresee the emergence of consciousness and self-awareness. The inherent novelty and layered complexity of emergent phenomena all but preclude prediction.

We are left, then, to ponder the possible existence of higher orders of emergence—stages of complexity that our brains can no more comprehend than a single neuron can comprehend the collective state of consciousness. Does the universe hold levels of emergence beyond individual consciousness, beyond the collective accomplishments of human societies? Might the cooperative awareness of billions of humans ultimately give rise to new collective phenomena as yet unimagined? If higher stages of emergence await our discovery, then science and theology may someday converge into a more unified vision of the cosmos and our place in it.

As we search beyond the cookbook "how" of life to questions of meaning and value, the concept of emergence holds a powerful message. Each of our lives is shaped by the same two competing powers of creation and destruction that have held sway throughout the history of the cosmos: emergence and entropy. The second law of thermodynamics states that entropy—the disorder of the universe—must increase. Yet in discrete, precious pockets of matter—on planets, in oceans, within our own conscious brains—astonishing levels of emergent complexity arise spontaneously.

This dramatic contrast provides a metaphor for our own lives. Some people choose the paths of hate, war, intolerance, destruction, and chaos to hasten the triumph of entropy—the dark side of the universe. By contrast, most people use their energies to foster emergence—to build cities, feed the hungry, create art, heal the sick, promote peace, and add to human joy and well-being in countless other ways, both large and small.

What awesome power each of us holds to do good or ill; a single cutting insult, a single winning smile. Perhaps therein lies life's meaning and value.

Notes

PREFACE

p. xiii   **It is possible:** A significant literature explores the contrasting views of life as a chance event (Monod 1971) *versus* a cosmic imperative (de Duve 1995a, Morowitz 2002).

p. xvi   **James Trefil:** Hazen and Trefil (1991), Trefil and Hazen (1992).

p. xvii   **"the unfolding of life . . . ":** Morowitz (2002, p. 84).

p. xvii   **conference in Modena, Italy:** The "Workshop on Life" was held September 3–8, 2000, as one of the satellite meetings before the Millennial World Meeting of University Professors in Rome, September 8–10. The conference proceedings are collected in Pályi et al. (2002).

PROLOGUE

p. 1   **This idea had received a boost:** Corliss et al. (1979, 1981). See Chapter 7 for more details on Jack Corliss's controversial claims.

p. 2   **Morowitz's dense tabulation:** The table of water's dielectric constant as a function of temperature and pressure came from Tödheide (1972, Table XI, p. 492). See also Uematsu and Franck (1981), Franck (1987), Shaw et al. (1991), and Franck and Weingartner (1999) for subsequent measurements by the same group at the Institute for Physical Chemistry and Electrochemistry, University of Karlsruhe. In spite of our surprise at seeing these results, experts in petroleum chemistry had long known about water's distinctive changes in properties at elevated temperature and pressure (e.g., Simoneit 1995). Theorist Everett Shock had incorporated these effects into his calculations of hydrothermal reactions relevant to prebiotic chemistry

(Shock 1990a, 1990b, 1992a, 1993; Shock et al. 1995). Nevertheless, few researchers in the origin-of-life community had made this connection, and no relevant experiments had been performed at high temperature and pressure.

p. 2   **detailed chemical scenario:** Wächtershäuser (1988a, 1990a, 1992). His work is reviewed in Chapters 8 and 15.

p. 3   **His name provided:** A perspective on Harold Morowitz's contributions to the founding of astrobiology is provided by Dick and Strick (2004, pp. 61-65).

p. 4   **Hat's pressure lab:** The apparatus and its operation is described in Yoder (1950).

p. 8   **"Humpane":** Jack Szostak writes, "It's the normal result to obtain a mess of hundreds of compounds. The central problem of prebiotic chemistry is how to avoid the universal tar of organic chemistry, and channel the chemistry into the products needed for the origin of life." [Jack Szostak to RMH, 21 August 2004]

1
THE MISSING LAW

p. 11   **"It is unlikely . . .":** J. H. Holland (1998, p. 3).

p. 11   **Two great laws:** Von Baeyer (1998) provides a history of the laws of thermodynamics.

p. 12   **The discovery of a dozen:** For a review of the principal laws of nature and their discovery, see Hazen and Trefil (1991).

p. 12   **scholars of the late nineteenth century:** The history of the idea that science has learned everything of significance appears in Horgan (1996), who defends and amplifies the idiotic claim.

p. 12   **Ilya Prigogine:** Prigogine's influential analyses of emergent systems, which he called "dissipative systems," appears in his books *Order Out of Chaos: Man's New Dialogue with Nature* (Prigogine 1984) and *Exploring Complexity: An Introduction* (Nicolis and Prigogine 1989). Of special interest to Prigogine were patterns that arise spontaneously in a shallow pan of boiling water (Bénard cells) and in certain types of slowly reacting chemicals (Belousov–Zhabotinski, or B–Z, systems). These systems, which could be analyzed with mathematical rigor, are representative of a larger class of phenomena in which energy flows through a collection of interacting particles. In the words of Wicken (1987, p. 5), "dissipation through structuring is an evolutionary first principle."

p. 13   **complex, turbulent convection:** The peculiar behavior of boiling water is addressed, for example, in Nicolis and Prigogine (1989, pp. 8-15).

p. 14   **patterns in water and sand:** An extensive technical literature analyzes the formation of sand patterning. Of special note are the classic works of Ralph Bagnold (1941, 1988).

p. 14   **understanding such simple systems:** Several reviewers question the idea that studies of patterning in simple mechanical systems such as sand can elucidate the behavior of much more complex biological systems. Graham Cairns-Smith writes: "I think that your discussion of emergent systems could do with a more explicit reference to the two main ways in which interestingly complex systems arise in biology—development and evolution. Development is modeled by, say, the Mandelbrot set: an amazing infinitely complex product with a childishly simple specification. And incidentally I don't see any new physical law here, just mathematics." (G. Cairns-Smith to RMH, 31 August 2004).

Jack Szostak echoes this opinion: "Part of the problem is the lack of a clear definition or understanding of what we mean by emergent phenomena, which leads to people referring to distinct things with the same term. So, in one sense, phenomena such as vortices or sand ripples are 'emergent' since they are collective phenomena not exhibited by the individual components of the system. On the other hand, there seem to be different kinds of phenomena one could also call 'emergent'.... Darwinian evolution emerges from the combination of replicating informational polymers and compartmentalization." [Jack Szostak to RMH, 21 August 2004] In other words, some scientists argue that emergent systems that become complex through a competitive selection process are fundamentally different from those that obey simple rules of interaction.

I'm more inclined to think that all emergent systems, whether shifting sand dunes or biological evolution, may ultimately be modeled based on a small set of "selection rules" reducible to mathematical statements. Ultimately, evolution through competitive selection represents simply another (though admittedly more elaborate) way that systems tend toward the most efficient way to dissipate energy.

p. 15   **A small band of scientists:** The Santa Fe Institute and its studies in emergence are discussed in Waldrop (1992) and Regis (2003).

p. 15   **John Holland:** Holland's influential works include *Hidden Order* (J. H. Holland 1995) and *Emergence: From Chaos to Order* (J. H. Holland 1998).

p. 15   **BOIDS:** This program is available at www.red3d.com/cwr/boids. The program tracks the flocking behavior of a hundred or so "BOIDS," each of which moves according to three rules: separation to avoid crowding flockmates, alignment to follow the flock's average direction, and cohesion to steer toward the average position of flockmates.

p. 15   **Physicist Stephen Wolfram:** Wolfram (2002).

p. 16   **Danish physicist Per Bak:** Bak's most accessible writings are found in his popular book, *How Nature Works: The Science of Self-Organized Criticality* (Bak 1996).

p. 16   **Santa Fe theorist Stuart Kauffman:** Kauffman (1993).

p. 16   **Nobel Laureate Murray Gell-Mann:** Gell-Mann and Tsallis (2004). Ideas of non-extensive entropy and related definitions of complexity are provided by Lopez-Ruiz et al. (1995), Shiner et al. (1999), Gell-Mann and Lloyd (2004), Latora and Marchiori (2004), Plastino et al. (2004). See also Gell-Mann (1994, 1995).

p. 17   **fossil-rich hundred-foot-tall cliffs:** The Miocene formations of Calvert County, Maryland, have been a Mecca for fossil collectors for two centuries. Details of the geology and paleontology are recorded in W. B. Clark (1904).

p. 17   **Factor 1:** The relationship between concentration of agents and complexity has the qualitative form of the so-called error function. This function begins at zero complexity for low concentrations of agents. As the

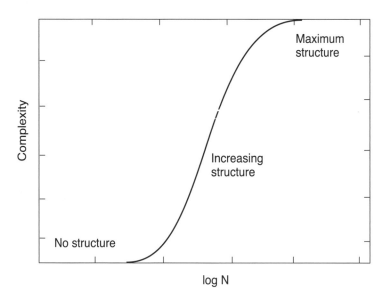

p. 17   The relationship between the concentration of interacting agents in a system (N) and the complexity of the system (C) has the qualitative form of the error function. At low N, no emergent structures arise, but as N increases, so does C, to an upper limit.

concentration rises, it reaches some critical value, and complexity begins to rise. Eventually, at a higher concentration of agents, the system's complexity achieves a maximum value.

p. 19   **One ant species:** Camazine et al. (2001, pp. 256-283). [Also E. O. Wilson to RMH, 9 April 2004; B. Fisher to RMH, 5 May 2004; C. W. Rettenmeyer to RMH, 12 May 2004]

p. 19   **Studies of termite colonies:** Solé and Goodwin (2000, p. 151).

p. 19   **spiral arm structure:** The formation of spiral arms requires "a central mass and a velocity large enough to produce a significant shear. . . . You also need time for the patterns to be established. The low mass objects don't survive long enough." [Vera Rubin to RMH, 7 April 2004]

p. 19   **Factor 2:** I suspect that one of the principal difficulties in quantifying emergent complexity is related to the varied ways that agents may be interconnected. The shifting interactions by immediate contacts of adjacent sand grains are not easily equated to the persistent chemical markers of moving ants or the elaborate networks of variable impulses that connect neurons. Nor are any of these examples exactly analogous to the interactions of molecules necessary for the origin of life.

p. 19   **A rounded grain:** Bagnold (1941, p. 85).

p. 20   **The conscious brain:** See, for example, Johnson (2001).

p. 20   **Factor 3:** The relationship between energy flow and complexity bears a qualitative similarity to a bell-shaped curve. At low energy flux, no patterning occurs and the complexity is zero. At some minimum energy flux, pattern formation begins and complexity quickly achieves a maximum value. Above a critical value, however, the energy flux is too great and the emergent patterns begin to disperse.

p. 20   **no pattern can emerge:** Physicist Paul C. W. Davies examines this issue from the standpoint of gravity, which is the initial and ultimate source of ordering in the universe (Davies 1999, p. 64).

p. 21   **Factor 4:** Cycling may be important in origin-of-life scenarios, but the imposition of any kind of cycle adds at least two new variables to an experiment. In the case of a temperature cycle, for example, one must select the duration of the cycle (typically on the order of minutes to days) and the two end-point temperatures. It may also be desirable to control the rate of temperature change (gradual versus abrupt), which adds additional variables. Needless to say, such added variables complicate an experiment and its interpretation.

p. 22   **amazing stone circles:** Kessler and Werner (2003).

p. 22 $\nabla E(t)$: This expression indicates both the flow of energy through the system, ∇E, and the cycling of that energy, which is a function of time, t.

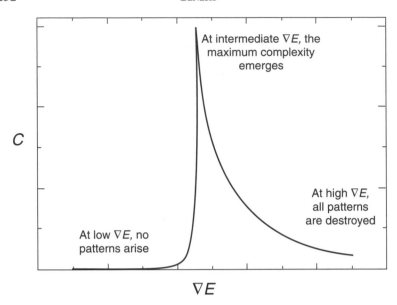

p. 20   The relationship between the energy flowing through a system of interacting agents (∇E) and the complexity of the system (C) has the form of a critical curve. At low ∇E, no emergent structures arise, but as ∇E increases above a critical value, structure rapidly appears and C increases. At high ∇E, however, patterns are destroyed.

2
WHAT IS LIFE?

p. 25   **"I know it when I see it.":** Associate Justice Potter Stewart of the U.S. Supreme Court made this statement as part of his concurring opinion in the 6-3 ruling that overturned the ban on pornographic films, June 22, 1964: "I shall not today attempt further to define the kinds of material . . . but I know it when I see it."

p. 25   **A recent origin-of-life text:** Lahav (1999, pp. 117-121).

p. 25   **"What is life?":** Pályi et al. (2002).

p. 26   **"top-down" approach:** Jack Szostak states "I don't think the earliest fossils tell us anything about life's earliest chemistry. These fossils are all quite sophisticated organisms." [Jack Szostak to RMH, 21 August 2004] Gustaf Arrhenius echoes this view: "The most 'primitive' organisms that we can lay our hands on are already hopelessly sophisticated biochemically; they are more like us than anything original." [Gustaf Arrhenius to RMH, 23 December 2004]

Indeed, a principal discovery of Precambrian paleontology is that modern-type cells have populated Earth for at least 80 percent of its history. The window for life's emergence is correspondingly brief.

p. 27 **"working definition":** See Joyce (1994).

p. 28 **thin molecular coating:** The idea of "flat life" has been explored by Wächtershäuser (1988a, 1992) among others. For more details see Chapter 15.

p. 28 **Claude Lévi-Strauss:** Lévi-Strauss, the founder of structural anthropology, argued that all humans share similar patterns of thought. His analysis of myths from various cultures, for example, stresses the common reliance on dichotomies in dealing with our role in the cosmos. See Lévi-Strauss (1969, 1978).

p. 28 **neptunists, . . . plutonists:** This debate, as well as that between catastrophists and uniformitarianists, was fueled by a conflict between scientific and religious interpretation of natural history. See, for example, Rudwick (1976) and Laudan (1987).

p. 29 **"The unfolding of life . . . ":** Morowitz (2002, p. 84). Several other authors, notably Christian de Duve (1995a), have adopted a similar framework of sequential episodes in life's emergence. See also Davies (1999), Maynard-Smith and Szathmáry (1999), and Bada (2004).

p. 30 **semantic question:** For a fuller explanation of these ideas, see Hazen (2002). Jack Szostak writes: "I used to think of the transition from non-life to life as a discontinuous event marked by the sudden start-up of Darwinian evolution. But as I've started to work on more aspects of the problem and look at it in more detail, what I see is more of a series of smaller steps." [Jack Szostak to RMH, 7 September 2004]

p. 30 **philosopher Carol Cleland:** Cleland and Chyba (2002).

p. 31 **Saturn's recently visited moon:** Evidence for volatiles at Titan's surface are presented by Lunine et al. (1999), Campbell et al. (2003), Griffith et al. (2003), and Lorenz et al. (2003). For the latest results from the Huygen's probe, see: http://saturn.jpl.nasa.gov/home/index.cfm.

3
LOOKING FOR LIFE

p. 33 **"Scientists turn reckless . . . ":** Lightman (1993, p. 41).

p. 33 **meteorite from Mars:** For an overview of Martian meteorites and their identification see McSween (1994). Goldsmith (1997) provides a popular account of the Allan Hills controversy.

p. 33 **Theorists maintained:** Calculations by Jay Melosh and co-workers of Caltech (Melosh 1988, 1993; Head et al. 2002) reveal that, although a lot of material is vaporized in large planetary impacts, rocks at the periphery of the impact site may be hurled unscathed off the surface into space. Such a process is thought to have transferred material from Mars to Earth. Similar impacts on Earth have certainly blasted terrestrial rocks into space, though the transfer of Earth rocks to Mars is much more difficult because of two gravitational impediments: Mars is much farther from the Sun, and it has a smaller gravitational field.

An alternative intriguing hypothesis is that cellular life may have arisen very early in Earth's history, and that some of these microbes may have been blasted into space by large asteroid impacts. Even if a subsequent globe-sterilizing giant impact occurred, life could have been reseeded by an Earth meteorite.

p. 33   **Allan Hills region of Antarctica:** The ice deserts of Antarctica are ideal hunting grounds for meteorites. These clean, dry regions persist for thousands of years, and dark-colored meteorites stand out starkly against their white surface. Scientists in helicopters or on snowmobiles can collect dozens of pristine specimens in a short field season. These areas are also valuable because they provide a relatively unbiased inventory of meteorite types. In more temperate regions, stony meteorites often blend into the surrounding rocky landscape, where they just weather away; only the resistant iron meteorites stand out, so they have always constituted the majority of meteorite finds. Antarctic collecting areas reveal that stony meteorites are much more common. For an informative visual overview, see the Web site of ANSMET, the Antarctic Search for Meteorites program, cosponsored by NASA and the Smithsonian Institution: http://geology.cwru.edu/~ansmet/.

p. 34   **A team of biologists:** D. S. McKay et al. (1996).

p. 34   **On August 7, 1996:** Dick and Strick (2004, pp. 179-201) provide a detailed history of this incident. The timing of the prepublication press conference was unusual because *Science* nearly always holds up press announcements until the afternoon before an article's publication date (in this case, Friday, August 16). However, the important story leaked more than a week before publication because, according to Goldsmith (1997), a top official in the Clinton administration told a prostitute, who, in turn, sold the information to a tabloid. *Science*, in agreement with the authors and NASA, therefore made the article available on their Web site eight days before print publication. [R. Brooks Hanson to RMH, 15 April 2004]

p. 35   **"Although there are alternative . . .":** In subsequent years, team leader David McKay elaborated on this argument by attempting to calculate the degree to which several weak lines of evidence coalesce to produce a strong probability when taken together. See, for example, D. S. McKay et al. (2002).

p. 36   **aggressively challenged:** *Science* received dozens of letters to the editor and technical comments challenging aspects of the D. S. McKay et al.

(1996) paper. Of these submissions, three letters were published in the September 20, 1996, issue and a technical comment appeared in the December 20, 1996, issue.

p. 36   **Point number one:** The most abundant organic molecules in the Allan Hills meteorite are polycyclic aromatic hydrocarbons, or PAHs, which are a ubiquitous component of carbon-rich meteorites as well as of soot, diesel exhaust, and myriad industrial processes. See, for example, Allamandola et al. (1985, 1989).

p. 36   **Point two:** An extensive and often contradictory literature has arisen around the question of Allan Hills' carbonate origins, particularly the temperature of their formation. See, for example, Harvey and McSween (1996) and Scott et al. (1997).

p. 36   **Skeptical experts:** The distinctive magnetite crystals, which possess a purity, crystal form, magnetic structure, and chainlike distribution characteristic of Earth microbes, provide the greatest remaining hope for proponents of the Martian life hypothesis. No plausible nonbiological explanation for these features has yet been proposed.

p. 36   **purported fossil microbes are too small:** Morowitz (1996). See also Schopf (1999, pp. 317-320).

p. 36   **The story became even more confused:** Steele et al. (2000a, 2000b).

p. 37   **J. William Schopf:** Schopf (1999, pp. 304-325).

p. 37   **"I was like Daniel ...":** Schopf (1999, p. 309).

p. 37   **"The minerals can't prove it ...":** Schopf (1999, p. 316).

p. 37   **"There are fine lines ...":** Schopf (1999, p. 325).

p. 38   **Such catastrophic events:** Maher and Stevenson (1988), Sleep et al. (1989). Estimates of the rate and magnitude of large impacts come largely from statistical studies of the sizes and ages of lunar craters.

p. 38   **We don't know:** Jack Szostak notes that "some people now think that a Hadean origin is likely, and that totally sterilizing impacts may have ended earlier than previously thought. And even following a sterilizing event, cells blasted into space might have reseeded the Earth." [Jack Szostak to RMH, 21 August 2004]

p. 38   **oldest known fossilized penis:** The original report was Siveter et al. (2003). The story appeared in *USA Today* on December 12, 2003. There was much scientific interest in this discovery because highly varied penile appendages constitute the basis for classifying the group of crustaceans called copepods.

p. 39   **Apex Chert:** Schopf (1993). Apex Chert fossils had been reported earlier, for example in Awramik et al. (1983) and in Schopf and Packer (1987). The 1993 paper was the first to enumerate 11 species of ancient Apex microbes.

Bruce Runnegar writes, "Although it is more dramatic to highlight the 1993 announcement of Schopf's discovery of Earth's oldest fossils, one should be aware that the same kind of announcement came out of the Schopf lab in the late 1970s from rocks of about the same age in the North Pole area of Western Australia. These were discovered by Stan Awramik at UC Santa Barbara and some people still believe these are the 'best of the oldest.' However, there are problems with these as well." [B. Runnegar to RMH, 4 March 2005]

p. 39   **world's leading experts:** Among Schopf's most noted works is the volume he edited on Precambrian paleontology, *Earth's Earliest Biosphere: Its Origin and Evolution* (Schopf 1983). In that work he presents his own review of the earliest known microbes (Schopf and Walter 1983).

p. 39   **UCLA protocol:** Schopf (1999, pp. 71-100).

p. 40   **Bonnie Packer:** [James Strick to RMH, 1 September 2004] The initial publication was Schopf and Packer (1987). Packer was the first person to observe unusual microstructures in Apex Chert samples.

p. 40   **Controversy erupted:** Brasier et al. (2002). An account of the controversy is presented by Knoll (2003, pp. 60-65).

p. 41   **Further study:** The geology of the Apex Chert site is still a matter of considerable debate. Brasier and Australian geologist John Lindsey highlighted these concerns at subsequent seminars, including a joint presentation at the Carnegie Institution's Geophysical Laboratory on July 1, 2002. For opposing viewpoints see Lindsay et al. (2003a, 2003b), Brasier et al. (2004), and Schopf et al. (2002).

p. 41   **"We reinterpret . . .":** Brasier et al. (2002, p. 77).

p. 41   **a rebuttal:** Schopf et al. (2002).

p. 42   **"News and Views":** Gee (2002). See also Dalton (2002).

p. 42   **This debate:** The Astrobiology Science Conference 2002 (the second such meeting) took place from April 7 to April 11 at NASA Ames Research Center. Abstracts of the meeting are available at http://web99.arc.nasa.gov/abscon2.

p. 42   **Schopf spoke:** The original NASA videotape of this debate was provided by Lynn Rothschild of the NASA Ames Research Center. A description of the session, including a photograph of Schopf peering at Brasier, appears in Dalton (2002).

p. 42   **soften his assertion:** While Schopf did back off his claim of fossil cyanobacteria in the lecture, he repeated the arguments in a later article, claiming that "Several of the Apex species seem almost indistinguishable from living cyanobacteria." (Schopf 2002, p. 173).

p. 43   **Raman spectroscopic data:** see also Pasteris and Wopenka (2003).

p. 44   **equally difficult to *disprove*:** Scientists are taught to evaluate alternative hypotheses based on Occam's razor: Accept the hypothesis that requires the fewest assumptions. But how does one choose between evidence for and against the existence of ancient microbial life, or life in a Martian meteorite, for that matter? If your philosophical bias is that life emerges rapidly and is thus ubiquitous in the universe, then any tantalizing sign of life—carbon smudges or scrappy fossils—will provide encouraging supporting evidence. If you think life rare or unique, then the tenuous character of the same data will prevail. Consequently, to prove either case—that an ancient rock does or does not hold evidence for life—is extraordinarily difficult. Brasier et al. were successful in casting doubt on Schopf's claims of microbial fossils, but they came nowhere close to disproving that the carbon residues were biogenic in origin.

p. 44 **Meanwhile, paleontologists:** Claims of 3.5-billion-year-old microfossils from South Africa were published by Furnes et al. (2004).

4

EARTH'S SMALLEST FOSSILS

p. 47 **"Millions of brutal years . . .":** Simpson (2003, p. 72).

p. 48 **Andrew Knoll:** The "Origin of Life" Gordon Research Conference took place from July 27 to 31, 1997. Two of Knoll's reviews (Knoll 1996, 2003) provide accessible overviews of Earth's earliest fossil cells.

p. 50 **severe analytical challenge:** Simultaneously with our efforts, Derek Briggs and co-workers at the University of Bristol produced similar electron microprobe maps of fossils (Orr et al. 1998).

p. 51 **Rhynie, Scotland:** The fossils of the Rhynie Chert are documented in the classic studies of Kidston and Lang (1917-1921).

p. 51 **Kevin Boyce:** Boyce's analytical work on Devonian plant fossils is described in Boyce et al. (2001, 2003).

p. 54 **Analytical studies:** Schidlowski et al. (1983), Schidlowski (1988).

p. 54 **mammoth bones:** Koch (1998).

p. 54 **fossil coal:** McRae et al. (1999).

p. 54 **Burgess Shale:** Butterfield (1990) reports bulk isotopic values of approximately −27 for the Burgess Shale. The classic 525-million-year-old Burgess Shale deposits of British Columbia preserve a diverse soft-bodied fauna in exquisite detail. These fossil forms are vividly described in Gould (1989).

p. 55 **Values as low as −50:** The distribution of ancient carbon isotope values has been tabulated, for example, by Schidlowski et al. (1983) and H. D. Holland (1997).

p. 55 **Walter's work:** See, for example, Walter et al. (1972), Walter (1976, 1983).

p. 55   **"It's Strelley Pool Chert . . .":** The age and geological setting of the Strelley Pool Chert has been described by C. P. Marshall et al. (2004) and Allwood et al. (2004).

p. 59   **Earth's Oldest "Fossils"?:** Bruce Runnegar writes, "I would prefer to title this section "Earth's Oldest Life" . . . . Fossils have been reported, even named (*Isuasphaera*), from the Isua rocks by Hans-Dieter Pflug but no one accepts them as remains of living organisms. What is being discussed here is only, at best, 'isotopic fossils.'" [B. Runnegar to RMH, 4 March 2005]

p. 59   **oldest known rocks:** Moorbath et al. (1986), Nutman et al. (1996, 1997).

p. 59   **Stephen J. Mojzsis:** Mojzsis et al. (1994, 1996). See also E. K. Wilson (1996) and H. D. Holland (1997) for analyses of the report.

Bruce Runnegar writes, "This is a dramatic way to present the discovery, but light carbon had been reported from the Isua rocks since the 1960s, I believe. Manfred Schidlowski, in particular, published many articles arguing that Isua provided evidence for life on Earth prior to the Mojzsis et al. article in *Nature*." [B. Runnegar to RMH, 4 March 2005]

p. 59   **Akilia rocks posed problems:** Fedo and Whitehouse (2002). See also Whitehouse (2000), Lepland et al. (2005). Kerr (2002a), Simpson (2003), and Dick and Strick (2004) provide analyses of the controversy. Aivo Lepland of the Geological Survey of Norway, who spent five years attempting to duplicate Mojzsis's results without success, has raised additional questions about the validity of the results. As a consequence, Gustaf Arrhenius, a coauthor on the 1996 paper, has distanced himself from the original conclusions (Lepland et al. 2005). "I think there must have been a mix-up of the samples," he says (Dalton 2004, p. 688).

More convincing evidence for ancient biogenic carbon is provided by Rosing (1999), who reported negative carbon isotope values from 3.7-billion-year-old sediments from the Isua region of Greenland. [Gustaf Arrhenius to RMH, 8 August 2004]

p. 60   **plausible explanations:** See, for example, van Zuilen et al. (2002). Negative carbon isotope fractionation is known to occur inorganically, for example, at high temperature, when iron carbonate breaks down to graphite plus iron oxide.

5
IDIOSYNCRASIES

p. 61   **"The ability of . . .":** Lahav (1999, p. 64).

p. 62   **industrial chemists:** The standard industrial process for synthesizing chainlike carbon molecules is called Fisher–Tropsch synthesis.

p. 62 **Life builds hydrocarbons:** The biochemical process for synthesizing chainlike molecules is detailed in Lehninger et al. (1993, pp. 642-649).

p. 62 **produced abundantly:** Allamandola et al. (1985, 1989, 1997), Allamandola and Hudgins (2003). PAHs are identified by their distinctive infrared emission spectra.

p. 63 **4-ring molecules called sterols:** Lehninger et al. (1993, pp. 669-674).

p. 64 **A crucial requirement:** For a survey of molecular biomarkers see K. E. Peters and Moldowan (1993).

p. 65 **The Hopane Story:** Hopanes were discovered by Guy Ourisson of the Université Louis Pasteur and co-workers (Rohmer et al. 1979, Ourisson and Albrecht 1992, Ourisson and Rohmer 1992, Tritz et al. 1999). The identification of hopane-related compounds in ancient Australian rocks is reported by Summons et al. (1999), with a commentary by Knoll (1999). For related work on ancient biomolecules from Australian rocks, see Summons and Walter (1990), Summons et al. (1996), and Buick et al. (1998).

p. 65 **In 1999:** Brocks et al. (1999).

p. 67 **2-methylhopanoid:** Bruce Runnegar writes, "As all extant cyanobacteria, so far as is known, make 2α-methylhopanes, the time of origin of this innovation may significantly predate the origin of the last common ancestor of living cyanobacteria." [B. Runnegar to RMH, 4 March 2005]

p. 67 **Biosignatures and Abiosignatures:** This section is adapted from Hazen et al. (2002).

p. 70 **biomolecular fragment:** See the Web site www.biocyc.org for a database that illustrates many of the hundreds of common small molecules found in cells.

p. 72 **microbial corrosion:** Steele et al. (1994).

p. 72 **Steele's assignment:** Steele et al. (2000a, 2000b).

p. 73 **the term "astrobiology":** Dick and Strick (2004, p. 205).

p. 73 **Enspel, Germany:** Toporski et al. (2002).

p. 75 **MASSE:** For information on the project development visit http://astrobiology.ciw.edu/main.php. The group proposes to look for morphological as well as chemical evidence (Cady et al. 2003). For an overview of Andrew Steele's life-detection experiments, see Whitfield (2004). The difficulties of detecting life remotely on Mars are underscored by the ambiguous results of the Viking mission in the 1970s (Levin and Straat 1976, 1981).

INTERLUDE—
GOD IN THE GAPS

p. 77 **Darwinists rarely mention the whale:** This quote comes from Alan Haywood's (1985) *Creation and Evolution.* These sentiments are echoed in Duane T. Gish's (1985) *Evolution: The Challenge of the Fossil Record,* in which he says: "There are simply no transitional forms in the fossil record between marine mammals and their supposed land mammal ancestors. . . . It is quite entertaining, starting with cows, pigs, or buffaloes, to attempt to visualize what the intermediates may have looked like. Starting with a cow, one could even imagine one line of descent which prematurely became extinct, due to what might be called an "udder failure" (pp. 78-79).

p. 78   **35-million-year-old** *Basilosaurus:* Gingerich et al. (1990).

p. 78   *Rodhocetus:* Gingerich et al. (1994).

p. 78   *Ambulocetus:* Thewissen et al. (1994).

p. 78   **new proto-whale species:** Thewissen et al. (2001). See also the companion commentary by Muizon (2001).

p. 80   **Michael Behe and William Dembski:** Behe (1996) and Dembski (1999, 2004). For opposing views see, for example, Pennock (2002) and Forrest and Gross (2004).

p. 80   **deeper problem:** K. R. Miller (1999) presents a compelling case against the idea of "God in the gaps."

6
STANLEY MILLER'S SPARK OF GENIUS

p. 83   **"The idea that . . .":** S. L. Miller (1953, p. 528). Günter Wächtershäuser writes: "Oparin, in fact, never suggested such an atmosphere, as can be verified by reading his books of 1924 and 1938" [Günter Wächtershäuser to RMH, 24 June 2004]

p. 83   **spontaneous generation:** The complex story of spontaneous generation, a theory that persisted throughout the nineteenth century, is described by Farley (1977), Fry (2000), and Strick (2000).

p. 83   **seventeenth-century invention:** The invention of the microscope was to biology what the invention of the telescope was to astronomy. Discoveries of microorganisms, the cell, and even smaller internal cellular structures transformed biology. See Ford (1985).

p. 84   **Lazzaro Spallanzani:** See, for example, Dolman (1975) for a biographical account and bibliographic citations.

p. 84   **Englishman John Needham:** Westbrook (1974) provides a biographical sketch and bibliographic sources.

p. 84   **Louis Pasteur:** Pasteur's motivation for these experiments in spontaneous generation is explored in Geison (1974), who also provides extensive bibliographic citations.

p. 85   **In 1871, Charles Darwin:** The letter is item 7471 in the Darwin online database: http://darwin.lib.cam.ac.uk. For a discussion of Darwin's views, see Fry (2000, pp. 54-57).

p. 85 **life requires liquid water:** See, however, the intriguing speculations of Steven Benner of the University of Florida (Benner 2002), who warns against "Earth-o-centrism." Perhaps, he notes, some other medium, such as liquid ammonia, might foster an alternative biochemistry on other worlds. Exploring this idea further, The Royal Society of London held a conference on "The Molecular Basis of Life: Is Life Possible Without Water?" December 3–4, 2003 (Ball 2004).

p. 86 **Alexander Oparin:** Oparin's 1924 work first appeared in English in 1938 (Oparin 1924, 1938), but it was not widely available to English-speaking audiences until Bernal (1967), which included a translation. According to Günter Wächtershäuser, the rarely cited book *Mechanische-Physiologische Theorie der Abstammungslehre* by Swiss botanist Carl Wilhelm von Nägeli (1884) includes a prescient description of "the origin of life in a broth of emerging, growing and evolving protein particles." [Günter Wächtershäuser to RMH, 24 June 2004]

p. 86 **"primordial soup":** This phrase follows J. B. S. Haldane's (1929) description of the early ocean as achieving "the consistency of hot dilute soup" (p. 247).

p. 86 **J. B. S. Haldane:** Haldane's choice of *The Rationalist Annual*, a periodical largely devoted to the promotion of rationalism and secular education, may seem an odd one for a theoretical paper on origin-of-life chemistry. Cooke (2004) chronicles the colorful history of the Rationalist Press Association, including Haldane's participation.

p. 87 **chemist Harold Urey:** Urey, a scientist of unusual breadth, won the 1934 Nobel Prize in chemistry for his discovery of deuterium, the heavy isotope of hydrogen and the essential component of "heavy water." He was also an authority on Earth's primitive atmosphere (Urey 1951, 1952), which led to his speculations about the prebiotic formation of organic compounds.

p. 87 **Jeffrey Bada:** Wills and Bada (2000).

p. 87 **Scientists revere simple:** Miller's original article (S. L. Miller 1953) contains a rather sketchy outline of the experiment. Additional details are provided by S. L. Miller (1955) and Wills and Bada (2000).

p. 90 **mid-February:** Wills and Bada (2000, p. 47) quote Stanley Miller's recollection of a mid-December 1953 submission. However, records in the Harold Clayton Urey papers (Scripps Institution of Oceanography Archives) include a manuscript receipt from *Science* dated February 16, 1953. [Antonio Lazcano to RMH, 30 August 2004; Jeffrey Bada to RMH, 1 September 2004]

p. 90 **Miller's first publication:** S. L. Miller (1953). The *New York Times* article, "Life and a glass Earth," appeared on May 17, 1953, page E10.

p. 90 **The Miller–Urey experiment:** Historical perspectives are provided by Wills and Bada (2000) and Bada and Lazcano (2003).

While Miller has received widespread acclaim for his experiment, some scientists and historians are less convinced of the originality of the Miller–Urey research. Similar experiments were conducted decades earlier by the German chemist Walter Löb (1906, 1914), who employed similar apparatus and also succeeded in synthesizing the amino acid glycine (see Mojzsis et al. 1998). Löb's research, however, was not designed to probe the chemistry of life's origins, nor was it meant to mimic prebiotic environments.

p. 90   **Independent confirmation:** Miller's experiments were repeated first by Hough and Rogers (1956) and Abelson (1956).

p. 90   **Walter Löb:** Löb (1906, 1914). Gustaf Arrhenius decries the lack of credit given to Löb. He states that the reverence accorded to the Chicago work is "an American myth originally based on cultural and linguistic ignorance, later on unwillingness to acknowledge the original work." [Gustaf Arrhenius to RMH, 26 December 2004] In addition, Löb died at a relatively early age and was thus unable to promote his findings.

p. 91   **John Oró:** Oró (1960, 1961a). Rensselaer Polytechnic Institute chemist James Ferris and co-workers elaborated on the role of HCN in prebiotic chemistry (Ferris et al. 1978).

p. 91   **Other chemists:** See Shapiro (1988) for a measured assessment of efforts to synthesize ribose by plausible prebiotic pathways.

p. 92   **Orgel and co-workers:** Sanchez et al. (1966). It is important to note that global dilution need not imply local dilution. Regarding this point, Louis Allamandola writes: "This is where I think an exogenic ice/ice residue has a great intrinsic advantage over endogenous processes, even given the total amounts are small with respect to a planetary reservoir. These ices and residues are not dilute." [Louis Allamandola to RMH, 6 July 2004]

p. 92   **the longest experiments:** The use of freezing to synthesize HCN polymers is related in Wills and Bada (2000, pp. 51-52).

p. 92   **by the 1960s:** The composition of the Archean Earth's atmosphere is a matter of significant debate. Few scientists today accept Miller's model atmosphere of methane, ammonia, and hydrogen. The majority view is that carbon dioxide and nitrogen were the dominant constituents of a relative unreactive atmosphere (H. D. Holland 1984; Walker 1986; Kasting 1990, 1993, 1994, 2001). Hiroshi Ohmoto of Pennsylvania State University, by contrast, has long argued that the early atmosphere featured significant oxygen content (Ohmoto et al. 1993, Ohmoto 1997). Tian et al. (2005) proposed an alternative hydrogen-rich atmosphere.

Nevertheless, it is likely that local pockets of reducing gases may have promoted organic synthesis. [Jack Szostak to RMH, 21 August 2004] Bada (2004) writes, "Even though reducing conditions may not have existed on a global scale, localized high concentrations of reduced gases may have existed around volcanic eruptions. . . . The localized release of reduced gases by volcanic eruptions on the early Earth would likely have been immediately exposed to intense lightning" (p. 6).

p. 93   **Miller and his supporters continue to counter:** There is a kind of logic to the argument that the early atmosphere must have been reducing because the resulting synthesis mimics biochemistry. Orgel (1998a, p. 491) states, "It is hard to believe that the ease with which sugars, amino acids, purines and pyrimidines are formed under reducing-atmosphere conditions is either a coincidence or a false clue planted by a malicious creator."

p. 93   **"If God did not ...":** as quoted in Wills and Bada (2000, p. 41).

p. 93   **extremely dilute solution:** The improbability of biochemical reactions arising from the dilute primordial soup has emerged as the central objection to the Miller hypothesis in the theories of Günter Wächtershäuser. In a dilute solution with hundreds or thousands of different solutes, the chance that a desired chemical reaction will occur between any two molecules is small. He states, "As far as I'm concerned, the soup theory is more of a myth than a theory, because it doesn't explain anything." (Hagmann 2002, p. 2007). For a more comprehensive critique, see Wächtershäuser (1994).

p. 93   **another nagging problem:** Stanley Miller himself often acknowledges this difficulty. In a 1992 *Discover* article, he said, "The first step, making the monomers, that's easy. We understand it pretty well. But then you have to make the first self-replicating polymers. . . . Nobody knows how it's done." (Radetsky 1992, p. 78). Miller repeated this refrain in a 1998 *Discover* article: "It's a problem. How do you make polymers? That's not so easy." (Radetsky 1998, p. 36).

7
HEAVEN OR HELL?

p. 95   **"It is we ...":** Gold (1999, p. v).

p. 96   **Metabolism:** For a useful overview of the surprising diversity of microbial metabolism, see Nealson (1997a).

p. 96   **Our view of life:** For a description of this research, see Radetsky (1992).

p. 97   **On this particular dive:** In 1979, scientists discovered that some of these vents spew out lots of dissolved minerals that precipitate in a thick black cloud as ocean and vent waters mix—a so-called "black smoker."

p. 97   **"Could the hydrothermal vents ...":** Jack Corliss as quoted in Radetsky (1992, p. 76).

p. 97   **others close to the story:** [John Baross to RMH, 24 June 1998 and 10 March 2004; Sarah Hoffman to RMH, 23 July 2004] Jack Corliss did not respond to requests for information. Hoffman provided a 28-page document with a detailed history of the development of the hydrothermal-origins hypothesis and its subsequent presentation at lectures and in print.

p. 98 **"ideal reactors . . .":** Corliss et al. (1981, p. 62).

p. 98 **much too hot:** Several papers from the Miller group focus on the supposed instability of amino acids under hydrothermal conditions, including S. L. Miller and Bada (1988), Bada et al. (1995), and Bada and Lazcano (2002). Other authors attempted to counter these arguments (Holm 1992).

p. 98 **". . . a real loser":** Stanley Miller as quoted in Radetsky (1992, p. 82).

p. 99 **Eventually the Corliss:** The Corliss et al. (1981) paper appeared in a supplementary section of papers presented at a symposium on the "Geology of the Oceans," which was part of the 26th International Geological Congress in Paris. A later paper was authored by Baross and Hoffman (1985).

p. 99 **John Baross remains active:** Much of John Baross's recent work focuses on barophilic (pressure-loving) and thermophilic (heat-loving) microbes. See, for example, Baross and Deming (1995).

p. 99 **Sarah Hoffman's graduate:** Corliss's abandonment of origins research was underscored when he delivered a lecture on his origin hypothesis, "Emergence of Living Systems in Archean Sea Floor Hot Springs," at the Geophysical Laboratory on January 8, 2001. The lecture offered no new insights beyond his work of the 1980s; indeed, he often interjected that "this is a 1986 lecture." He seemed to forget many of the details of his model—temperatures, depths, organic chemistry—and his answers to several questions were vague and uninformative.

p. 100 **Everywhere they looked:** Dozens of recent books and articles document microbes in extreme environments (Madigan and Marrs 1997, Wharton 2002), including Antarctic ice (Price 2000, Thomas and Dieckmann 2002), boiling hot springs (Stetter et al. 1990, Hoffman 2001), acidic pools and streams (Zettler et al. 2002), deep-ocean hydrothermal zones (Pedersen 1993), and crustal rocks (Krumholz et al. 1997, Chapelle et al. 2002).

Concurrent with discoveries of abundant deep microbes were the findings of molecular biologist Carl Woese (Woese and Fox 1977; Woese 1978, 1987). Woese applied the techniques of molecular phylogeny to construct a tree of life (see Chapter 10). He discovered that the traditional divisions of life into five kingdoms was incorrect and that the most primitive living cells (i.e., microbes deeply rooted in the evolutionary tree of life) are extremophiles that live in hydrothermal conditions (Pace 1997). This result suggested to some researchers that the first life-forms might have been similar extremophiles. Such a conclusion is not certain, however, because life might have arisen in a cooler surface environment and subsequently radiated into extreme environments. A large impact might then have killed off all surface organisms, leaving extremophiles as our last common ancestors.

p. 100   **Savannah River:** Frederickson and Onstott (1996) provide a popular account of this research.

p. 100   **loaded with microbes:** The Savannah River samples from a depth of 400 meters support from 100 to 10 million microbes per gram of rock. By comparison, a typical gram of topsoil might hold a billion microbes per gram.

p. 101   **Subsequent drilling studies:** Parkes et al. (1993), Stevens and McKinley (1995), Krumholz et al. (1997), Pedersen et al. (1997), Chapelle et al. (2002), and D'Hondt et al. (2002, 2004).

p. 101   **Tullis Onstott:** Frederickson et al. (1997), Colwell et al. (1997), Tseng and Onstott (1998) and Onstott et al. (1998). For popular accounts of this research, see Frederickson and Onstott (1996), Monastersky (1997), and Kerr (2002b).

p. 101   **"It was 'Don't . . .' ":** Schultz (1999, p. 1).

p. 102   **Thomas Gold:** Bondi (2004).

p. 103   **In 1977:** Gold (1977). The first peer-reviewed publication of these ideas appeared two years later (Gold 1979). See also Gold and Soter (1980).

p. 104   **Siljan Ring:** Gold (1999, pp. 105-123).

p. 104   **Seven years:** The controversy was summarized for *Science* by reporter Richard Kerr (1990). Additional points of view are provided by Donofrio (2003) and by reviewers of Gold's book (Brown 1999, Margulis 1999, Parkes 1999, Von Damm 1999). Brown's review in *American Scientist*, "Upwelling of Hot Gas," is particularly contemptuous. Few authors seem to have considered the possibility of a middle ground. Might hydrocarbons arise from both surface life and from deep sources? After all, there's a lot of carbon, and hydrocarbons happen.

p. 104   **"the deep hot biosphere":** Gold (1992, 1997, 1999).

p. 104   **invited Gold:** The seminar, entitled "The Deep Hot Biosphere," took place on April 28, 1998.

p. 105   **Tommy Gold helped:** Gold died on June 22, 2004, two weeks after suffering a massive heart attack. On June 7, 2004, he had sent me a preliminary review of the first half of this book. "The only comment I want to make before I have read it all very carefully is that you refer too often to the ocean vents," he wrote. He argued that many other deep environments also contribute organic molecules and might have been more conducive to the origin of life. "But more about all this when I have read with some care what you sent me." Sadly, that addendum never came. [Thomas Gold to RMH, 7 June 2004]

p. 105   **Heaven Versus Hell:** A version of this text appeared in Hazen (1998).

p. 106   **ancient Mars or Venus:** Speculations on the habitability of various bodies in the solar system include Reynolds et al. (1983), Sagan et al. (1992), Thompson and Sagan (1992), and Boston et al. (1992).

8
UNDER PRESSURE

p. 107   **"Where is this . . .":** Wächtershäuser (1988a, p. 480).

p. 108   **Geophysical Laboratory:** High-pressure research at the Geophysical Laboratory is described in Hazen (1993) and Yoder (2004). For a general history of the Carnegie Institution, see Trefil and Hazen (2002).

p. 108   **well funded by NASA:** NASA's Astrobiology Institute (NAI) was founded in 1998 with the Carnegie Institution as one of 11 charter research teams. Additional groups were added in 2001 and 2003. For more information, see the NAI Web site: http://nai.arc.nasa.gov.

p. 108   **Our first experiments:** Cody et al. (2000).

p. 108   **In a later set of experiments:** Brandes et al. (1998).

p. 108   **Carrying on with this line:** Brandes et al. (2000).

p. 109   **"This proposal is based . . .":** S. L. Miller and Bada (1988, p. 609).

p. 109   **"a real loser":** Stanley Miller as quoted in Radetsky (1992, p. 82).

p. 110   **again and again:** These quotes appear in Wills and Bada (2000, pp. 98 and 101). "Ventists," annoyed at this condescending label, have been known to refer to Bada and colleagues as "Millerites"—or, after a couple of beers, "Miller lites." In fact, the name "ventists" was coined by RPI chemist James Ferris, who also dubbed Miller and his followers "arcists" (as quoted in Simoneit 1995, p. 133).

p. 111   **roles as varied:** The possible roles of minerals in life's origin are reviewed by Hazen (2001).

p. 111   **"Before enzymes...":** Wächtershäuser's ideas initially appeared in four papers (Wächtershäuser 1988a, 1988b, 1990a, 1990b). The central tenets were summarized in "Pyrite formation, the first energy source for life: A hypothesis" (Wächtershäuser 1988b), which emphasizes the role that Popperian philosophy played in the theoretical effort. That paper was submitted to *Systematic Applied Microbiology* in March of 1988 and published later that year. A more elaborate presentation, "Before enzymes and templates: Theory of surface metabolism" (Wächtershäuser 1988a) soon followed in *Microbiology Review*. The paper that first brought his ideas to the attention of a wide audience appeared two years later in the *Proceedings of the National Academy of Sciences* (Wächtershäuser 1990a). Articles submitted to the *Proceedings* are often communicated by an Academy member; Wächtershäuser's paper, "Evolution of the first metabolic cycles," was sponsored by Karl Popper himself. The full-blown theory is articulated in the massive "Groundworks for an evolutionary biochemistry: The iron–sulfur world" (Wächtershäuser 1992). Several subsequent papers clarify and elaborate on the model and respond to a growing barrage of comment and criticism (Wächtershäuser 1993, 1994, 1997).

p. 111   **Karl Popper:** Popper's key ideas are summarized in three of his books, *The Logic of Scientific Discovery* (Popper 1959), *Conjectures and Refutations: The Growth of Scientific Knowledge* (Popper 1963), and *Objective Knowledge: An Evolutionary Approach* (Popper 1972). Wächtershäuser's presentation of his own hypothesis in terms of "Theory Darwinism" (essentially, competition among rival hypotheses and survival of the fittest) is most clearly presented in Wächtershäuser (1988a).

p. 111   **"During breakfast . . ."**: As quoted in Nicholas Wade, "Gunter Wachtershauser: Amateur Shakes up on Recipe for Life" (*New York Times*, April 22, 1997).

p. 112   **patched together a theory:** A number of widely cited origin-of-life hypotheses, including those based on a primordial soup, can be criticized for their poor predictive ability.

p. 113   **"You don't mind . . ."**: Günter Wächtershäuser as quoted in Radetsky (1998, p. 36). Experiments designed to test aspects of Wächtershäuser's theory include Blöchl et al. (1992), Keller et al. (1994), Huber and Wächtershäuser (1997, 1998), and Huber et al. (2003).

p. 113   **"It takes maybe two weeks":** Günter Wächtershäuser in response to Robert Hazen at his Geophysical Laboratory seminar, March 23, 1998. Other estimates vary widely. De Duve (1995b, p. 428) suggests, "millennia or centuries, perhaps even less." See also Lazcano and Miller (1994) and Fry (2000, pp. 125-126).

p. 113   **"not a new idea":** Bada and Lazcano (2002, p. 1983). The paper they refer to, "A note on the origin of life" (Ycas 1955), introduces the idea of an autocatalytic cycle of metabolites as the first living system. Dick and Strick (2004, p. 64) agree that "Ycas had pioneered the 'metabolism first' idea." However, Noam Lahav notes that Alexander's (1948) discussion of autocatalysis predates that of Ycas. [Noam Lahav to RMH, 27 August 2004]

p. 114   **January 1998:** "We have not met, but your work of the past decade on the role of sulfides in organic synthesis and the origin of life is having a profound effect on our current research," I wrote. "We would be delighted if you could schedule a trip to the States, for which we would pay expenses." [RMH to Günter Wächtershäuser, 14 January 1998]

p. 114   **Wächtershäuser delivered his lecture:** The Geophysical Laboratory seminar, "Chemoautotrophic Origin of Life in an Iron–Nickel–Sulfur World," took place on March 23, 1998.

p. 114   **"We would like to explore . . .":** [RMH to Günter Wächtershäuser, 8 April 1998]

p. 115   **I was left:** That memorable meeting was the last contact any of us had with Günter Wächtershäuser until Harold Morowitz received a stern letter on Wächtershäuser & Hartz legal stationary dated October 11, 2000. Copies of the letter were also sent to the bosses of everyone involved, including Bruce Alberts, president of the National Academy of Sciences; Maxine Singer, president of the Carnegie Institution; and Alan Merten, president of George Mason University. Harold Morowitz, with three coauthors (including George Cody), had recently published a short article in the *Proceedings of the National Academy of Sciences* on life's most primitive metabolic cycle. Morowitz et al. (2000) proposed that molecules used in the so-called "reductive citric acid cycle" (what I refer to in the text as the "reverse citric acid cycle") are highly selected, with a long list of distinctive features. Such selectivity, Harold concluded, suggests that life's earliest metabolism is deterministic and likely to be the same on any planet or moon where life emerges. Wächtershäuser denounced the article as improperly claiming credit for ideas that were originally his, and he demanded an immediate retraction. Harold ignored the veiled threat of legal action, but we were saddened that a brilliant man with such creativity and vision could so remove himself from the cooperative spirit of scientific research.

p. 115   **His first paper:** Brandes et al. (1998).

p. 115   **He followed up:** Brandes et al. (2000).

p. 115   **In spite of these successes:** Bada et al. (1995). The hydrothermal stability of amino acids is also discussed by Shock (1990b), Hennet et al. (1992), and W. L. Marshall (1994).

p. 116   **a few centuries:** Wolfenden and Snider (2001).

p. 116   **preserve proteins:** Recent studies on the essential bone protein osteocalcin underscores the ability of minerals to stabilize organic molecules. Hoang et al. (2003) document the complex structure of osteocalcin and illustrate how it binds strongly to hydroxyapatite, the principal mineral constituent of bone. This binding not only provides strength and flexibility to bone but also protects osteocalcin from the rapid decay experienced by most other proteins. Christine Nielsen-Marsh and co-workers at the University of Newcastle (Nielsen-Marsh et al. 2002) exploit this feature to extract and sequence osteocalcin from fossil bison bones more than 50,000 years old. Over time, osteocalcin undergoes slight mutations in its amino acid sequence. Comparison of small differences in this sequence among fossils of various species and ages thus reveals patterns of mammalian evolution.

Bruce Runnegar writes, "I'm skeptical of the report of collagen in dinosaur bone. Osteocalcin maybe, but I wait to be convinced about collagen." [B. Runnegar to RMH, 4 March 2005]

p. 116   **Additional evidence:** Lemke et al. (2002) and Ross et al. (2002a, 2002b).

p. 117   **molecules might separate out:** The idea of the separation from water of an amino-acid-rich phase first appeared in von Nägeli (1884). This idea was also a central feature of Oparin's original hypothesis (Oparin 1924, 1938), as well as several subsequent proposals. See, for example, Fox and Harada (1958) and Fox (1965, 1988).

p. 117   **more work to be done:** In spite of these results, David Ross emphasizes that amino acids cannot survive long enough in a hydrothermal environment to start life. "The utility of hydrothermal work is that it allows us to accelerate reactions to convenient times so that we can study them. No more than that. . . . The key reactions leading to life involved very, very slow reactions. Half-lives of a million years or more would be the order of the day, and it would take a graduate student of unusual longevity, durability, and endurance to get any data on such reactions." [David Ross to RMH, 14 July 2004]

p. 118   **Such reactions occur rapidly:** These experiments are detailed in Cody et al. (2004).

p. 118   **Fischer–Tropsch (F–T) synthesis:** A number of researchers have studied F–T synthesis under hydrothermal conditions (McCollom and Simoneit 1999, McCollom and Seewald 2001, Foustoukos and Seyfried 2004).

p. 118   **Recent intriguing analyses:** Sherwood-Lollar et al. (1993, 2002).

p. 119   **thiols and thioesters:** The possible role of these sulfur-containing compounds in life's origins has been detailed by de Duve (1995a, 1995b).

9
PRODUCTIVE ENVIRONMENTS

p. 121   **"The limits of life . . .":** Nealson (1997b, p. 23677). Nealson notes that the quote appeared "again in slightly different form in 1998 in a Caltech lecture, and in the final form in Nealson and Conrad (1999)."

p. 121   **immense tenuous clouds:** Ehrenfreund and Charnley (2000) enumerate several types of interstellar structures where organic synthesis occurs. The density of these structures ranges from about one atom per cubic centimeter in "interstellar clouds" to a million atoms per cubic centimeter in "dense molecular clouds." Louis Allamandola writes: "A volume roughly the size of a large auditorium is home to only one tenth-micron-sized dust grain. It is the dust which absorbs background starlight, making dark interstellar molecular clouds dark. Thus their sizes are enormous, often measured in thousands of light-years." He notes that in these clouds, PAHs, which are produced primarily during earlier star-formation processes, are more abundant than all other interstellar molecules combined. [Louis Allamandola to RMH, 16 July 2004]

p. 122   **more than 140 different compounds:** Infrared spectroscopy of molecular clouds is reviewed by Pendleton and Chiar (1997) and Rawls (2002). An up-to-date list of all identified interstellar molecular species is available at: http://www-691.gsfc.nasa.gov/cosmic.ice.lab/interstellar.htm.

p. 122   **Allamandola and co-workers' experiments:** See, for example, Bernstein et al. (2002). A lively, accessible, and richly illustrated account of this research appears in *Scientific American* (Bernstein et al. 1999a). Similar experiments have been performed by a research team based at the Leiden Observatory in The Netherlands (Muñoz Caro et al. 2002).

James Ferris writes: "[Allamandola] spent a good part of his career working in Mayo Greenberg's lab in Leiden before going to Ames. Greenberg developed this apparatus and approach and Allamandola adopted the design after going to Ames." [James Ferris to RMH, 22 August 2004]

p. 123   **Evidence from space:** Reviews of the rich variety of organic molecules recovered from meteorites include Cronin and Chang (1993), Glavin et al. (1999), Becker et al. (1999), Ehrenfreund and Charnley (2000), and Cody et al. (2001a). Cometary organic molecules, though less well documented, are described by Chyba et al. (1990) and Ehrenfreund and Charnley (2000). Kwok (2004) also emphasizes the important role of "proto-planetary nebulae"—the envelopes of gas and dust around newly forming stars—in the production of organic molecules. See also Oró (1961b), Urey (1966), Kvenvolden et al. (1970), Cronin and Pizzarello (1983), Anders (1989), Cronin (1989), Delsemme (1991), Engel and Macko (1997), Irvine (1998), and Pizzarello and Cronin (2000).

p. 123   **seeded abundantly:** Allamandola argues that interstellar ice particles could have provided much more than simple organic building blocks. "These could well have been a source of prebiotic/biogenic molecules which played a specific role in the origin of life. Going even further, I am beginning to think we should also consider the possibility that the chemistry in these ices might be even more advanced, perhaps being a fountainhead of life" [Louis Allamandola to RMH, 16 July 2004]

p. 123   **"that's garbage...":** Stanley Miller as quoted by Radetsky (1992, p. 80).

p. 123   **"Even if cosmic debris . . .":** Jeffrey Bada as quoted by Radetsky (1998, p. 37). More recently Bada (2004, p. 7) has softened his objections and points to a combination of sources: "It is now generally assumed that the inventory of organic compounds on the early Earth would have been derived from a combination of both direct Earth-based syntheses and inputs from space."

p. 123   **It's hard to imagine:** In the mid-1990s, when NASA scientists subjected carbon-rich meteorite fragments to realistic impact velocities of 3 miles per second, about 99.9 percent of the amino acids were obliterated (Peterson et al. 1997). They concluded that impact velocities of the Murchison and other amino-acid-bearing meteorites must have been significantly less, perhaps owing to aerobraking, thus preserving more of the delicate organic molecules.

p. 123   **impacts don't destroy all:** Other shock experiments to induce organic synthesis have been reported by C. P. McKay and Borucki (1997), who used a high-energy infrared YAG laser to shock-heat a gaseous sample to temperatures greater than 10,000°C. These experiments simulate the effects of an impact on the atmosphere.

p. 123   **giant experimental gas gun:** See Blank et al. (2001).

p. 124   **idea of Friedemann Freund:** Freund et al. (1980, 1999, 2001).

p. 124   **"Maybe," he remarked:** Friedemann Freund to Wesley Huntress, undated note ca. 2000 attached to a copy of Freund et al. (1999).

p. 125   **"I am a hundred percent sure . . .":** [Anne M. Hofmeister to RMH, 21 November 2002]

p. 125   **synthetic magnesium oxide:** Crystal growth of MgO is described by Freund et al. (1999), who also document the identity of extracted carboxylic acids. Freund's studies on MgO properties include Kathrein and Freund (1983), Kötz et al. (1983), and Freund et al. (1983).

p. 126   **gem-quality olivine:** Freund's olivine samples come from the classic San Carlos, New Mexico, locality, which is an active gem-producing area. The gemmy green olivine crystals (also known as peridot), as much as an inch across, comprise up to 50 percent of the basaltic rock, which was formed deep in the crust. Samples are widely available commercially, but the outcrops occur on the San Carlos Indian reservation, so access and collecting is restricted.

p. 126   **100 parts per million carbon:** Keppler et al. (2003) disagree with this claim for San Carlos olivine. Their experiments on olivine growth under carbon-saturated conditions produced crystals with no more than 0.5 part per million carbon.

p. 126   **"I ran a sample . . .":** [George R. Rossman to Anne M. Hofmeister, 20 December 2002] Many mineralogists and solid-state chemists discount Freund's findings as experimental artifacts resulting from surface contamination. See, for example, Keppler et al. (2003) and M. Wilson (2003).

p. 127   **no single dominant source:** A few authors imply that it remains a mystery which of several sources of organic molecules—Miller-type surface synthesis, hydrothermal processes, impacts, or deep-space synthesis—was dominant. Orgel (1998a, p. 491), for example, states: "Three popular hypotheses attempt to explain the origin of prebiotic molecules: synthesis in a reducing atmosphere, input in meteorites and synthesis on metal sulfides in deep-sea vents. It is not possible to decide which is correct." Similarly, Miller and his colleagues have at times discounted both hydrothermal zones and extraterrestrial sources as trivial (S. L. Miller and Bada 1988; Stanley Miller and Jeffrey Bada as quoted in Radetsky 1992, 1998). Other more ecumenical estimations of multiple organic sources include Chyba and Sagan (1992) and Lahav (1999).

INTERLUDE—
MYTHOS VERSUS LOGOS

p. 129   **"People of the past . . .":** Armstrong (2000, p. xiii). She continues, "Myth looked back to the origins of life, to the foundations of culture, and to the deepest levels of the human mind. Myth was not concerned with practical matters, but with meaning." By contrast, "*Logos* was the rational, pragmatic, and scientific thought." Armstrong argues, "People of Europe and America [have] achieved such astonishing success in science and technology that they began to think that *logos* was the only means to truth and began to discount *mythos* as false and superstitious."

p. 129   **"Whoa, . . .":** [Margaret H. Hazen to RMH, 30 May 2004]

10
THE MACROMOLECULES OF LIFE

p. 133   **"To purify . . .":** Lehninger et al. (1993, p. 5).

p. 133   **One of the transforming discoveries:** Lehninger et al. (1993). The extraordinary Web site http://biocyc.org/ECOO157/new-image?object= Compounds tabulates all known small organic molecules from the microbe *Escherichia coli.*

p. 134   **"I can no longer . . .":** Friedrich Wöhler to his teacher Jacob Berzelius, February 28, 1828. This discovery (Wöhler 1828) was made in the same year that Wöhler isolated and named the element beryllium.

p. 134   **Four key types of molecules:** For an overview of the characteristics of sugars, amino acids, carbohydrates, and nucleic acids, see Lehninger et al. (1993).

p. 135   **Sugars are the basic building blocks:** Estimates of Earth's total biomass place cellulose, the abundant glucose polymer that forms leaves, stems, trunks, and other plant support structures, at the top of the list (Lehninger et al. 1993, pp. 298 et seq).

p. 135   **For every useful molecule:** Of the several proposed prebiotic mechanisms for biomolecular synthesis, Miller's original spark experiments produce perhaps the highest percentage of useful molecules. Fully 6 percent of the carbon atoms introduced as CH_4 in some of his experiments were incorporated into amino acids, for example (S. L. Miller and Urey 1959a). For this reason alone, Miller and his supporters often argue that electric discharge in a reducing atmosphere is the most likely origin scenario.

p. 136   **life is even choosier:** See, for example, Bonner (1995). The problem of chiral selection is discussed in more detail in Chapter 13.

p. 136   **molecular phylogeny:** See, for example, Pace (1997), Pennisi (1998), and Sogin et al. (1999).

p. 137   *The Canterbury Tales:* Barbrook et al. (1998). Similar techniques of textual comparison have long been employed for shorter manuscripts, but this study used the same computer algorithms as applied to genomic data.

p. 138   **Carl Woese:** The original proposal for three domains of life, including the Archaea, appears in Woese and Fox (1977). See also Woese (1978, 1987, 2000, 2002). A biographical sketch of Carl Woese, including an overview of his work, is provided by Morell (1997).

p. 139   **thrive at elevated temperature:** Perspectives on the proposition that the last common ancestor was an extremophile are provided, for example, by Gogarten-Boekels et al. (1995), Forterre (1996), and Reysenbach et al. (1999).

Bruce Runnegar writes, "There is a last common ancestor and it was a highly derived organism. It tells us very little about Earth's earliest cells in the same way that living birds do not reveal the attributes of dinosaurs." [B. Runnegar to RMH, 4 March 2005]

p. 141   **swap sections of DNA:** Gogarten et al. (1999), Doolittle (2000), and Woese (2002).

p. 141   **"last common ancestor":** See, for example, Woese (1998) and Ellington (1999). An important conclusion of recent studies is that, because of gene transfer, there is no single last common cellular ancestor. Woese (1998, p. 6854) writes: "The universal ancestor is not a discrete entity. It is, rather, a diverse community of cells that survives as a biological unit. The universal ancestor has a physical history but not a genealogical one."

p. 141   *all cells employ RNA:* Woese (2002).

p. 141   **simple metabolic strategy:** Woese (1998, p. 6855) states that the biochemical repertoire of the universal ancestor included "a complete tricarboxylic acid cycle, polysaccharide metabolism, both sulfur oxidation and reduction, and nitrogen fixation." Pace (1997, p. 734) comments that "the earliest life was based on inorganic nutrition."

p. 142   **primordial "oil slick":** Lasaga et al. (1971). See also Morowitz (1992).

11
ISOLATION

p. 143   **"The self-assembly process . . .":** Deamer (2003, p. 21).

p. 143   **Lipid molecules:** For an accessible overview of lipid molecules and their spontaneous organization into bilayers, see Tanford (1978) and Segré et al. (2001).

p. 144   **Alec Bangham:** See, for example, Bangham et al. (1965). Some researchers initially called these structures "banghasomes." [Harold Morowitz to RMH, 10 August 2004]

p. 144   **Luisi and co-workers:** Luisi (1989, 2004), Luisi and Varela (1989), Luisi et al. (1994), Bachmann et al. (1992), and Szostak et al. (2001). See also Segré et al. (2001).

p. 145   **counted as classics:** Pasteur (1848), Miller (1953), and Bernstein et al. (1999b).

p. 146   **Deamer returned:** Deamer and Pashley (1989). For additional information, see Zimmer (1993) and Deamer and Fleischaker (1994).

p. 146   **Murchison meteorite:** For a description of the Murchison meteorite and related research, see Grady (2000, pp. 350-352).

p. 147   **Their straightforward procedure:** The eclectic mix of organic molecules in Murchison included some species, like amino acids, that were soluble in water; some, like lipids, that were soluble in chloroform or other organic solvents; and a complex tarry residue, called by the generic name "kerogen," which is difficult to analyze. Recent studies by Cody et al. (2001a) suggest that this residue consists of a complexly linked mass of rings, chains, and other smaller groups of atoms. It is not evident that such insoluble matter could have played much of a role in prebiotic chemistry.

p. 148   **breakthrough moment:** The discovery paper by Deamer and Pashley (1989) was entitled "Amphiphilic components of the Murchison carbonaceous chondrite: Surface properties and membrane formation." In this article they state, "If amphiphilic substances derived from meteoric infall and chemical evolution were available on the prebiotic earth following condensation of oceans, it follows that surface films would have been present at air-water interfaces. . . . This material would thereby be concentrated for self-assembly into boundary structures with barrier properties relevant to function as early membranes."(p. 37) This paper was especially noteworthy because it followed by a year the publication of a theoretical paper by Morowitz et al. (1988) that proposed such an origin scenario.

p. 148   **NASA Ames team:** Dworkin et al. (2001).

p. 149   **a colorful photograph:** *The Washington Post* (Kathy Sawyer, "IN SPACE; CLUES TO THE SEEDS OF LIFE," January 30, 2001, p. A1).

p. 149   **astrobiology meetings:** The First Astrobiology Science Conference was held April 3–5, 2000, at the NASA Ames Research Center, Moffett Field, California. Deamer's lecture was entitled "Self-assembled Vesicles of Monocarboxylic Acids and Alcohols: A Model Membrane System for Early Cellular Life" (Apel et al. 2000).

p. 151   **we had made bilayer membranes:** These results were reported at the 221st Annual Meeting of the American Chemical Society, held in San Diego, California, April 1-5, 2001.

p. 151   **Recent work:** Knauth (1998) provides estimates of higher salinity in the Archean ocean. Salt inhibition of amphiphile self-organization is reported in Monnard et al. (2002).

p. 152   **atmospheric aerosols:** Dobson et al. (2000). See also Ellison et al. (1999), Tuck (2002), and Donaldson et al. (2004). These studies, which present theoretical analyses of aerosol dynamics and atmospheric residence times, build on earlier speculative comments regarding the possible roles of aerosols by Woese (1978) and Lerman (1986, 1994a, 1994b, 1996). Regarding Lerman's contributions, James Ferris writes: "Unfortunately a head injury in an automobile accident had a major effect on his life and he was unable to get a full paper written on this proposal. He discussed this proposal at meetings and it was well known in the origins of life community." [James Ferris to RMH, 22 August 2004].

12
MINERALS TO THE RESCUE

p. 155   **"But I happen to know . . .":** Updike (1986, pp. 328-329).

p. 155   **The first living entity:** Portions of this chapter were adapted from Hazen (2001).

p. 156   **Mineralogist Joseph V. Smith:** J. V. Smith (1998), Parsons et al. (1998), and J. V. Smith et al. (1999). Other authors, including Cairns-Smith et al. (1992), have also proposed that porous minerals might have provided a measure of protection for proto-life.

p. 157   **a primitive slick:** The oil-slick hypothesis was championed by Morowitz (1992) in his influential book *The Beginnings of Cellular Life: Metabolism Recapitulates Biogenesis.* See also Lasaga et al. (1971), who estimated that a primordial oil slick on the Archean ocean could have achieved a thickness of 1 to 10 meters.

p. 157   **British biophysicist John Desmond Bernal:** Bernal (1949, 1951). The Swiss-born geochemist Victor Goldschmidt also suggested that minerals played a role in life's origin, but his thoughts, presented as a lecture in 1945 and published posthumously (Goldschmidt 1952), had little impact on the origins community (Lahav 1999, p. 250).

p. 157   **In a 1978 study:** Lahav et al. (1978). See also Lahav and Chang (1976) and Lahav (1994).

p. 157   **NASA-sponsored teams:** Among the chemists who have studied roles of clays and other fine-grained minerals in prebiotic processes, two NASA Specialized Center of Research and Training (NSCORT) groups at Scripps Institution of Oceanography (La Jolla, California) and Rensselaer Polytechnic Institute (Troy, New York) have made notable contributions.

p. 157   **James Ferris:** Reports by Ferris and colleagues on mineral-induced polymerization of RNA, principally by the common clay montmorillonite and the phosphate hydroxyapatite, include Ferris (1993, 1999), Holm et al. (1993), Ferris and Ertem (1992, 1993), Ferris et al. (1996), and Ertem and Ferris (1996, 1997). Images of organic molecules on ideally smooth mineral surfaces have been published, for example, by Sowerby et al. (1996) and Uchihashi et al. (1999).

p. 157   **"activated" RNA:** Ferris writes: "My experiments work only if *activated* nucleotides are reacted. The thermodynamics is against self-condensation of nucleotides to form the phosphodiester bond in aqueous solution. That's why nature uses ATP in place of AMP to form RNA. By the way, ATP and ADP do not work in the clay catalyzed reaction so we use the imidazole activating group that was introduced first by other workers and popularized by Lohrmann and Orgel." [James Ferris to RMH, 22 August 2004]

p. 158   **Leslie Orgel:** Experiments on polypeptide formation are described in Ferris et al. (1996), Hill et al. (1998), and Liu and Orgel (1998).

p. 158   **"polymerization on the rocks":** Orgel (1998b). See also Acevedo and Orgel (1986).

p. 158   **One possible answer:** Chen et al. (2004). Subsequent work by Szostak's group reveals that a wide variety of powdered minerals promotes similar vesicle formation.

p. 159   **Gustaf Arrhenius:** Arrhenius and co-workers' studies of double-layer hydroxides appear in Arrhenius et al. (1993), Gedulin and Arrhenius (1994), and Pitsch et al. (1995).

p. 160   **Joseph Smith:** Smith's mineralogical proposal on "Biochemical evolution" appeared in J. V. Smith (1998), Parsons et al. (1998), and J. V. Smith et al. (1999).

p. 160   **A. G. (Graham) Cairns-Smith:** The clay-life hypothesis first appeared in "The structure of the primitive gene and the prospect of generating life" (manuscript dated October 1964). The first publication was Cairns-Smith (1968); see also Cairns-Smith (1977). Important book-length elaborations include *Genetic Takeover and the Mineral Origins of Life* (Cairns-Smith 1982) and *Clay Minerals and the Origin of Life* (Cairns-Smith and Hartman 1986). Popular accounts of these ideas include *Seven Clues to the Origin of Life* (Cairns-Smith 1985a) and a *Scientific American* article, "Clays and the origin of life" (Cairns-Smith 1985b).

p. 160　**"I believe . . . "**: Cairns-Smith (1985b, p. 900).

p. 160　**"Evolution did not . . . "**: Cairns-Smith (1985a, p. 107).

p. 161　**"The answer"**: Cairns-Smith (1985b, pp. 91-92).

p. 161　**"I'm an organic chemist . . ."**: A. G. Cairns-Smith seminar, "Clay Minerals and the Origin of Life," Carnegie Institution, June 16, 2003.

p. 162　**clay minerals commonly display:** Varieties of clay defects are illustrated in Cairns-Smith (1988, 2001).

p. 162　**"In two-dimensional . . . "**: Cairns-Smith (1985b, p. 96).

p. 163　**particularly stable sequences:** Cairns-Smith writes: "The word stable sounds like thermodynamically most stable whereas in fact any informational structure has to be at least a little bit unstable. Like any genetic information it would owe its prevalence to indirect effects that favour its own survival and/or propagation—e.g., suppose that a particular defect arrangement catalyses (a little bit) the production of di- or tri-carboxylic acids, which in turn assist clay synthesis by transporting aluminum." [A. Graham Cairns-Smith to RMH, 31 August 2004]

p. 164　**In 1988:** Cairns-Smith (1988).

p. 164　**"The first step . . ."**: Cairns-Smith (1988, p. 244).

p. 164　**"Can the material . . ."**: [A. Graham Cairns-Smith to RMH, 18 December 2003]

13
LEFT AND RIGHT

p. 167　**"Assemblage on corresponding . . ."**: Goldschmidt (1952, p. 101).

p. 167　**One of the stages:** A vast literature addresses the origin of biological chirality. Comprehensive reviews have been presented by Bonner (1991, 1995).

p. 169　**global excess:** Meticulous analyses of amino acids in some meteorites have revealed a small but significant excess of L-amino acids (Cronin and Pizzarello 1983, Engel and Macko 1997, and Pizzarello and Cronin 2000).

p. 169 **Louis Pasteur:** Pasteur (1848).

p. 169 **polarized light:** Numerous recent articles explore this idea (S. Clark 1999, Podlech 1999, and Bailey et al. 1998).

p. 169 **parity violations:** See, for example, Salam (1991).

pp. 169-170 **local, as opposed to global:** Some authors claim that an important philosophical distinction exists between deterministic global models of life (i.e., that some intrinsic feature of the universe demands left-handed amino acids) versus a chance local selection of left or right (i.e., life might have formed either way through stochastic processes). Note, however, that of the many symmetry-breaking models proposed, parity violations in beta decay provide the only truly universal chiral influence. Most authors conclude that this effect is so small as to be negligible in any realistic calculations of chiral selection (Bonner 1991, 1995). All other proposed symmetry-breaking mechanisms are local, though at vastly different scales (i.e., Popa 1997). Circularly polarized light from rapidly rotating neutron stars, for example, may selectively break down right-handed amino acids in one substantial volume of galactic space but will have the opposite effect in other volumes. Even if such a scenario led to a preponderance of left-handed amino acids in our region of the galaxy, an equal volume of space would have featured an excess of right-handed molecules.

p. 170 **local environments abounded:** Goldschmidt (1952), Lahav (1999), and Hazen (2004).

p. 171 **Albert Eschenmoser:** Eschenmoser (1994, 2004) and Bolli et al. (1997).

p. 173 **For most of the twentieth century:** For example, Ferris and Ertem (1992, 1993), Arrhenius et al. (1993), Gedulin and Arrhenius (1994), Pitsch et al. (1995), Ferris et al. (1996), Ertem and Ferris (1996, 1997), Hill et al. (1998), Liu and Orgel (1998), J. V. Smith (1998), Parsons et al. (1998), and J.V. Smith et al. (1999). See Chapter 12.

p. 173 *three* **separate points:** See, for example, Davankov (1997).

p. 173 **By the 1930s:** Tsuchida et al. (1935) and Karagounis and Coumonlos (1938). More recent work by Bonner et al. (1975) casts doubt on these accounts.

p. 173 **experiments were flawed:** For a more extensive discussion see Hazen and Sholl (2003) and Hazen (2004).

p. 174 **Edward Dana's *A Textbook:*** Dana (1958).

p. 181 **Glenn's research:** The study of amino acid racemization was pioneered by Ed Hare (Hare and Mitterer 1967, 1969; Hare and Abelson 1968) of the Carnegie Institution's Geophysical Lab (with whom Goodfriend worked for many years) and Jeffrey Bada (Bada et al. 1970, Bada and Schroeder 1972, and Bada 1972) of the Scripps Institution of Oceanography.

Hare and Bada developed a bitter rivalry, fueled by disagreements over priority in this research.

p. 181   **But his biggest and boldest:** Goodfriend and Gould (1997). See also Goodfriend et al. (2003).

p. 183   **aspartic acid had to be chemically modified:** Goodfriend (1991).

p. 184   **We wrote up the results:** Hazen et al. (2001). This episode highlights a recurrent problem in science: When does an experiment end (Galison 1987)? We had collected sufficient data, replicated on four crystals, and performed with duplicate runs and analyses, to provide statistically meaningful conclusions. This proof of concept therefore constituted a "publishable unit."

p. 185   **visit from Steve Gould:** Gould's (2002) mammoth book appeared in March 2002 to much notice and grudging admiration. In spite of his cancer, he returned to D.C. in April 2002 for book signings.

INTERLUDE—
WHERE ARE THE WOMEN?

p. 187   **"Where are the women?":** [Sara Seager to RMH, 14 November 2004]

14
WHEELS WITHIN WHEELS

p. 191   **"The origin of metabolism . . .":** Dyson (1985, 1999).

p. 191   **cosmic imperative.** de Duve (1995a).

p. 192   **which one came first:** The dichotomy between metabolism-first and genetics-first models is discussed by Lahav (1999, p. 189 et seq.) and Wills and Bada (2000, pp. 137-139). For varied viewpoints, see, for example, Dyson (1999), Morowitz (1992), de Duve (1995a), Orgel (1998a), and E. Smith and Morowitz (2004).

p. 192   **Those who favor genetics:** E. Smith and Morowitz (2004, p. 21) note: "One of the striking sociological features of biology today is the extraordinary importance placed on the sequencing and interpretation of DNA."

p. 193   **Self-Replicating Molecules:** For a general overview of self-replicating molecules, see E. K. Wilson (1998).

p. 194   **Julius Rebek, Jr.:** Much of Rebek's work on self-complementary molecules was performed while he was Camille Dreyfus Professor of Chemistry at MIT. This work is described in Tjivikua et al. (1990); Rebek (1994, 2002); Conn and Rebek (1994); and Wintner et al. (1994).

p. 194   **Reza Ghadiri:** Lee et al. (1996). See also Kauffman (1996) and Yao et al. (1997).

p. 194   **Self-complementary strands:** Von Kiedrowski (1986). See also Sievers and von Kiedrowski (1994) and Li and Nicolaou (1994).

p. 196   **"It has not escaped...":** Watson and Crick (1953, p. 737).

p. 196   **Stuart Kauffman:** Many of Kauffman's ideas are summarized in two books, *The Origins of Order* (Kauffman 1993) and *At Home in the Universe* (Kauffman 1995).

p. 197   **"autocatalytic networks":** An important theoretical treatment of the evolution of autocatalytic systems, called "the hypercycle," has been developed by Manfred Eigen of the Max Planck Institute for Biophysical Chemistry (Eigen 1971; Eigen and Schuster 1977, 1978a, 1978b, 1979). For a useful analysis, see Fry (2000, pp. 100-111).

p. 197   **a certain degree of sloppiness:** Darwinian natural selection is predicated on a certain degree of random variation in the characteristics of individuals, which leads to competition and selection. Maynard-Smith and Szathmáry (1999, p. 7) state: "Since, almost inevitably, one cycle would be more efficient in utilizing resources of the environment than the other, one would be 'naturally selected.'"

p. 198   **"That's for the chemists...":** As quoted by Harold Morowitz to RMH, ca. 2001. This attitude prompted John Maynard-Smith to refer to Kauffman's work as "fact-free science" (Davies 1999, p. 141).

p. 199   **An unbroken chemical history:** The principle of continuity— or congruity, as some call it—must apply to any origin-of-life scenario. Each increase in emergent complexity must arise in an unbroken sequence from the chemical processes of previous steps.

p. 199   **The Protenoid World:** Fox and Harada (1958), Fox and Dose (1977), and Fox (1956, 1960, 1965, 1968, 1980, 1984, 1988).

p. 199   **"Fox," Morgan would often remark:** As quoted in Dick and Strick (2004, p. 40).

p. 200   **Fox's career thrived:** Dick and Strick (2004, pp. 31-43), recount Fox's career and detail NASA's grant support of his work commencing in 1960.

p. 200   **"alive in some...":** Dick and Strick (2004, p. 41).

p. 200   **As early as 1959:** S. L. Miller and Urey (1959b). This reply was in response to a letter by Fox (1959).

p. 201   **Protenoid World was influential:** For objective analyses of Fox's influence, see Fry (2000, pp. 83-88) and Dick and Strick (2004, pp. 31-43). Fry concludes, "Though major parts of Fox's theory were later challenged by many researchers, his influence at the time was instrumental in turning the problem of the origin of life into a scientific subject." (p. 88)

p. 201   **nonrandom and deterministic:** Following the Miller–Urey experiments, the prevailing attitude favored models of origins by random chemical processes (Wald 1954).

p. 201   **marginalized:** Wills and Bada (2000, pp. 52-55) recount the Fox story: "Over time, he became more and more of a maverick in the field. Sadly, at the Eleventh International Conference on the Origin of Life, held in Orleans, France, in 1996, he was reduced to having placards of protenoid microspheres paraded around in the manner of a cartoon sandwich man predicting the Second Coming." (p. 55)

p. 201   **The Thioester World:** Christian de Duve's hypothesis is articulated in a series of books and articles, including an accessible presentation for general readers, *Vital Dust: Life as a Cosmic Imperative* (de Duve 1995a). See also de Duve (1991, 1995b).

p. 202   **a "volcanic setting":** de Duve (1995b, p. 435).

p. 202   **carbon–sulfur bond:** de Duve (1995a, 1995b) notes that the energy of the carbon–sulfur bond in thioesters is comparable to that of the phosphate bond in modern energy-rich molecules such as ATP. De Duve selects thioesters in his model because they are more plausible prebiotic molecules from a geochemical perspective.

p. 202   **steady supply:** Some experimental evidence supports the assumption of a steady supply of thioesters. Huber and Wächtershäuser (1997) produced thioesters of acetic acid in experiments that mimicked hydrothermal conditions, while Weber (1995) described the production of amino acid thioesters under similar conditions.

15
THE IRON–SULFUR WORLD

p. 205   **"You don't mind . . .":** Günter Wächtershäuser as quoted in Radetsky (1998, p. 36).

p. 205   **Günter Wächtershäuser's:** His theory is detailed in a series of papers, including Wächtershäuser (1988a, 1988b, 1990a, 1990b, 1991, 1992, 1993, 1994, 1997). For a comprehensive overview of Wächtershäuser's chemical ideas, along with those of other advocates of sulfide-driven prebiotic processes, see Cody (2004).

p. 205   **The contrast between heterotrophic and autotrophic:** For discussions of the heterotroph-first versus autotroph-first debate, see Lazcano and Miller (1996), Lahav (1999), Wills and Bada (2000), and E. Smith and Morowitz (2004). Note that the autotroph-first position is also, by necessity, deterministic; and it represents a metabolism-first viewpoint.

A measure of the relative complexity of autotrophs and heterotrophs is provided by the minimum number of genes required for cells to survive. Morowitz et al. (2004) estimate that heterotrophic cells, which obtain all essential molecules from their surroundings, require a minimum of approximately 500 genes. By contrast, modern autotrophic cells require at least 1,500 genes. This sharp contrast in genomic complexity is perhaps the strongest argument in favor of a heterotrophic origin.

p. 206   **irrelevant to the origin of life:** Wächtershäuser (1988b, p. 453) states the case: "My theory contrasts sharply with the ingenious prebiotic broth theory of Darwin, Oparin, and Haldane for I deny the preexistence of any arsenal of organic building blocks for life (such as amino acids). Rather, I assume the concentration of dissolved organic constituents in the water phase is negligible."

pp. 206-207   **precious little evidence:** A number of theoretical studies, notably by Everett Shock (then at Washington University in St. Louis), lent support to the idea of hydrothermal organic synthesis in general, if not the Iron–Sulfur World hypothesis in detail (e.g., Shock 1990a, 1990b, 1992a, 1992b, 1993; and Shock et al. 1995). Shock (now a professor at Arizona State University) and his students use thermodynamic models based on the relative energies of chemical products and reactants. They demonstrate that the inherent lack of equilibrium between oxygen-poor hydrothermal fluids and more oxidized seawater can drive metabolism and the formation of carbon–carbon bonds. For example, Shock et al. (1995, p. 141) write, "The amount of energy available was more than enough for organic synthesis from CO_2 or CO, and/or polymer formation, indicating that the vicinity of hydrothermal systems at the sea floor was an ideal location for the emergence of the first chemolithoautotrophic metabolic systems." (A "chemolithoautotroph" is an autotroph that gets its energy from the chemical disequilibrium of minerals.) Our experimental group was greatly influenced by Shock's efforts, and for a time he was a NASA Astrobiology Institute co-investigator with the Carnegie Institution team. By contrast, Wächtershäuser, whose publications are typically characterized by copious references to other work, has rarely cited Shock's papers.

p. 207   **the initial tests:** Drobner et al. (1990). A key aspect of this experiment was the exclusion of any oxygen, which might have poisoned the reaction by forming iron oxides. Hence, they describe "the formation of both pyrite and molecular hydrogen under fastidiously anaerobic conditions in the aqueous system of FeS and H_2S" (p. 742). See also Blöchl et al. (1992). This group also studied amino acid polymerization (Keller et al. 1994).

p. 207   **Wolgang Heinen and Anne Marie Lauwers:** Heinen and Lauwers (1996).

p. 207   **Subsequent experiments:** Huber and Wächtershäuser (1997, 1998) and Huber et al. (2003).

p. 207   **Our research group:** Cody et al. (2000, 2004). See also the commentary by Wächtershäuser (2000).

p. 208   **The Reverse Citric Acid Cycle:** Wächtershäuser (e.g., 1992, pp. 129 et seq.) proposes that the reductive citric acid cycle is the basis for the first autocatalytic cycle. Harold Morowitz has elaborated on this idea (Morowitz et al. 2000, Smith and Morowitz 2004). Others are not persuaded, however. Orgel (1998a, p. 495) notes that in the absence of enzymes, "the chance of closing a cycle of reactions as complicated as the reverse citric acid cycle, for example, is negligible."

As one possible counterargument, Harold Morowitz now advocates the role of "small molecule catalysts" such as the amino acid proline, "which can act as a catalyst in aldol condensations. Small molecule catalysis can then act as a self-organizing principle in forming metabolic networks." [Quoted from an announcement of Morowitz's lecture, "A Principle of Biochemical Organization: The Roots of Genetic Code Within the Intermediary Metabolism of Autotrophs," delivered at the Krasnow Institute for Advanced Study, George Mason University, 13 September 2004]

p. 208   **a simple philosophy:** These ideas are detailed in Morowitz (1992), in which he argues that "Metabolism recapitulates biogenesis."

p. 208   **In the mid-1960s:** Evans et al. (1966).

p. 209   **At a recent seminar:** Morowitz's seminar entitled "The Feed-Down Principle" was delivered at the Krasnow Institute for Advanced Study, George Mason University, on February 2, 2004. His theme (from the abstract): "Biology appears to be organized in a hierarchical fashion with a biochemical core consisting of the tricarboxylic acid cycle (TCA cycle or reductive TCA cycle) and the reaction network producing all of the key biochemical building blocks."

p. 210   **modern metabolic enzymes:** See, for example, Adams (1992) and Beinert et al. (1997).

p. 210   **formation of pyrite:** In Wächtershäuser's Iron–Sulfur World, the mineral pyrite plays a crucial role as the solid surface to which life clings. Under many geochemical conditions, pyrite has a positively charged surface, whereas the essential compounds in the reductive citric acid cycle are negatively charged molecules. The metabolic molecules thus bond strongly to the mineral surface—an essential characteristic of surface life. Otherwise, the metabolites would break free and be lost forever in the dilute ocean.

In this regard, a word about nomenclature is in order. The molecules that form the citric acid cycle are all acids in their electrically neutral state: acetic acid, pyruvic acid, oxaloacetic acid, and so on. In solution, however, these molecules typically give up a hydrogen atom to become negatively charged molecules called acetate, pyruvate, oxaloacetate, and so on. For this reason the citric acid cycle is sometimes called the citrate cycle.

p. 211   **What do experiments:** Cody et al. (2001b).

p. 212   **Cody found evidence:** Cody et al. (2001b); see also Cody

(2004). They observe the reaction of 1-nonene (a 9-carbon molecule) with nickel sulfide in a formic acid solution to 10-carbon carboxylic acids. This reaction implies that a similar beta-pathway reaction of 3-carbon propene to form 4-carbon methyl acrylic acid is possible.

p. 212   **And perhaps someday:** The lack of experimental verification of many chemical steps in the Iron–Sulfur World hypothesis remains a point of criticism (de Duve and Miller 1991). In particular, the relative instability of oxaloacetate, which tends to decarboxylate rapidly, poses a significant challenge to the development of a reductive citric acid cycle before enzymes (Lazcano and Miller 1996).

p. 213   **Michael Russell and Allan Hall:** Russell and Hall (Russell et al. 1993, 1994; Russell and Hall 1997, 2002). An intriguing feature of the Russell and Hall model is its reliance on mineral phases with iron and/or nickel in more than one valence state: Mackinawite $[(Fe,Ni)_{1+x}S]$ is proposed as the principal membrane sulfide, whereas greigite (Fe_3S_4) and violarite $(FeNi_2S_4)$ serve the role of catalysts. The structures of these minerals bear striking resemblances to the structures of Fe–Ni–S clusters at the core of key metabolic enzymes, including ferredoxins and CO dehydrogenase (Russell and Hall 2002). See also Adams (1992), Beinert et al. (1997), Rawls (2000), Doukov et al. (2002), and J. W. Peters (2002) for descriptions of a variety of metal–sulfur clusters in enzymes.

p. 213   **"flat life":** Wächtershäuser's Iron–Sulfur World hypothesis, for example, relies on two-dimensional growth across a pyrite surface. A number of other authors have speculated on the possibility of flat life, including E. Smith and Morowitz (2004). For a critique of this proposal, see de Duve and Miller (1991) and Wills and Bada (2000, p. 138).

16
THE RNA WORLD

p. 215   **"It is generally believed . . .":** Joyce (1991, p. 391).

p. 215   **classic 1968 paper:** Orgel (1968, p. 381) writes, "It is argued that the evolution of the genetic apparatus must have required the abiotic formation of macromolecules capable of residue-by-residue replication. This suggests that polynucleotides were present even in the most primitive ancestors of contemporary organisms."

p. 216   **a messy business:** Fox and Harada (1958).

p. 216   **"I must confess . . .":** Orgel (1986, p. 127).

p. 216   **RNA ribozymes:** Early work on ribozymes includes Kruger et al. (1982), Guerrier-Takada et al. (1983), Bass and Cech (1984), Zaug and Cech (1986), Been and Cech (1988), and Altman et al. (1989). Subsequent work has produced ribozymes that accomplish a wide range of catalytic function (see, e.g., Cech 1990, 1993; Noller et al. 1992).

In fact, the theoretical papers of Woese (1967), Crick (1968), and Orgel (1968) all anticipated the discovery of ribozymes by postulating the existence of genetic molecules that act as catalysts.

p. 217   **RNA World:** The term "RNA World" was proposed in a short note by Harvard chemist Walter Gilbert (1986), in the same year that Alberts (1986) and Lazcano (1986) proposed that catalytic RNA preceded DNA. Gilbert postulated the RNA World as a step in life's development when RNA molecules were sufficient "to carry out all the chemical reactions necessary for the first cellular structures." These ideas received much subsequent elaboration, for example by Orgel (1986), Joyce et al. (1987), Joyce (1989, 1991, 1996), Biebricher et al. (1993), Lehman and Joyce (1993), and Joyce and Orgel (1993).

p. 217   **study of ribosomes:** Ban et al. (2000) and Nissen et al. (2000). Nobelist Thomas Cech (2000) contributed a high-profile analysis of this work, entitled "The ribosome is a ribozyme."

Jack Szostak writes: "You could write a book on just the intrigue and controversy and personalities involved in determining the structure of the ribosome—arguably the most important 'molecular fossil' of modern cells (all right, at least comparable to the citric acid cycle)." [Jack Szostak to RMH, 21 August 2004].

p. 218   **called coenzymes:** Lehninger et al. (1993) provide an overview of the functions of coenzyme-A and other coenzymes that incorporate nucleotides.

p. 218   **"riboswitches":** Winkler et al. (2002, 2004) and Sudarsan et al. (2003).

p. 218   **uses rather simple molecules:** e.g., Lehninger et al. (1993).

p. 219   **RNA nucleotides, by contrast:** RNA nucleotides are themselves assembled from three smaller molecules: the 5-carbon sugar ribose, one of four different cyclic compounds called bases, and an orthophosphate (PO_4) group. Bases have been synthesized in a variety of plausible prebiotic experiments, including S. L. Miller's (1953) original experiments and in many subsequent studies (Oró 1960, 1961a; Sanchez et al. 1967; Ferris et al. 1978; Robertson and Miller 1995; Shapiro 1995; and Hill and Orgel 2002). For an excellent overview of this progress, see S. L. Miller and Lazcano (2002, pp. 92-100).

The synthesis of the 5-carbon sugar ribose, a relatively unstable molecule, is more difficult, though it has been produced in modest yields by the reaction of formaldehyde—the so-called formose reaction (Butlerow 1861, Reid and Orgel 1967, Cairns-Smith et al. 1972, Shapiro 1988, and Schwartz and de Graaf 1993). Interestingly, Ricardo et al. (2004) report that borate minerals may stabilize ribose preferentially, and Springsteen and Joyce (2004) find that ribose passes through lipid membranes an order of magnitude more easily than other 5-carbon sugars. These effects may have enhanced the prebiotic selection of ribose.

Equally problematic is a reliable prebiotic source of orthophosphate (Weber 1982, Westheimer 1987, Yamagata et al. 1991, de Graaf et al. 1995, Lazcano and Miller 1996, Kornberg et al. 1999, Kornberg and Fraley 2000, and de Graaf and Schwartz 2000).

Even more daunting is a viable mechanism to assemble the three components—base, sugar, and phosphate—into a nucleotide. Pitsch et al. (1995) reported successful synthesis of sugar phosphates, but ribose and bases are, in general, not stable under the same set of conditions (Joyce 1989, Joyce and Orgel 1993, Lahav 1999, and Zubay and Mui 2001). Wills and Bada (2000, p. 131) conclude: "It is very difficult to imagine how even the building blocks of RNA could have arisen by themselves, much less chains of RNA constructed from the building blocks."

p. 219   **catalytic RNA sequences:** Even if individual nucleotides could be synthesized in abundance, there is no known way to induce them to polymerize in a realistic prebiotic environment. A number of experiments have demonstrated the spontaneous formation of nucleotide–nucleotide bonds, for example, in the presence of clay minerals (Ferris et al. 1988, 1996; Ferris and Ertem 1992, 1993; and Ertem and Ferris 1996, 1997), but these nucleotides were "activated"—chemically altered to enhance their reactivity.

p. 219   **Biologists seem reasonably confident:** An extensive half-century of literature, commencing with the work of George Gamow (1954), considers the origin of the DNA-Protein world and the genetic code. This topic postdates the chemical origin of life and thus is beyond the scope of this book. For reviews, see de Duve (1995a, Chapter 6), Lahav (1999, Chapter 17), Maynard-Smith and Szathmáry (1999, Chapter 4), and Wills and Bada (2000, Chapter 7). Hayes (2004) provides a concise survey of the origin of the genetic code.

p. 219   **intractable gap:** Leslie Orgel (2003, p. 213) states, "I believe that it is very unlikely that RNA did arise prebiotically on the primitive Earth." S. L. Miller (1997, p. 167) echoes that belief: "RNA is an unlikely candidate."

17
THE PRE-RNA WORLD

p. 221   **"I've been waiting . . .":** [Simon Nicholas Platts to RMH, 27 May 2004]

p. 221   **"Identifying the first . . .":** S. L. Miller (1997, p. 167).

p. 221 **Albert Eschenmoser:** Eschenmoser's first studies (Eschenmoser 1991, 1993, 1994, 1999; Bolli et al. 1997) focused on 6-carbon (or hexose) sugars, which assemble into polymers called pyranosyls (he called these alternative nucleic acids p-RNAs). In 2000, his group reported the surprising synthesis of TNA nucleic acid with the 4-carbon sugar threose (Schöning et al. 2000). Leslie Orgel (2000, p. 1307) writes: "The existence of a molecule that is significantly 'simpler' than RNA, that resembles RNA more closely than do peptide nucleic acids . . . is encouraging to those who believe that RNA was preceded by one or more simpler genetic materials." For useful overviews, see Eschenmoser (1999, 2004).

Regarding this work, Jack Szostak writes: "Eschenmoser [is] one of the world's pre-eminent organic chemists, who should have shared the Nobel Prize with Woodward for conformational analysis and B12 synthesis—who in his retirement has taken on the task of synthesizing systematically all the reasonable alternatives to RNA in order to answer the question of why nature chose the nucleic acids that are used today in all cellular life." [Jack Szostak to RMH, 21 August 2004]

p. 222 **"peptide nucleic acid":** The concept of PNAs was introduced by Nielsen et al. (1991), Egholm et al. (1992), and Nielsen (1993), and elaborated on by many other researchers (Wittung et al. 1994, Diederichsen 1996, Diederichsen and Schmitt 1998, Nielsen 1999, and Orgel 2003). Stanley Miller (1997, p. 169) comments, "PNA has demonstrated that nucleic acids with backbones other than sugar phosphates need to be considered."

The original PNA used a nonbiological amino acid called aminoethyl glycine—a nonchiral molecule. Subsequent studies found that PNA cannot form a DNA-like double strand if the peptide backbone is constructed from homochiral biological amino acids (known as α amino acids). Rather, the backbone must consist either of so-called β amino acids or of strictly alternating D and L α amino acids (Diederichsen 1996, Diederichsen and Schmitt 1998, and Orgel 2003).

p. 222 **plausible genetic molecules:** Steven Benner takes the chemical range of RNA-like molecules even further as he imagines the possibilities of alternate biochemistries (Switzer et al. 1989, Piccirilli et al. 1990, Bain et al. 1992, and Hutter and Benner 2003). He has developed an Artificially Expanded Genetic Information System (AEGIS) with eight new base pairs and a variety of new genetic polymers, some of which might be stable in nonaqueous liquids, such as ammonia, and thus serve as models for alien biochemistries (Benner and Hutter 2002). Miller and co-workers also explored the possibility of alternative bases (Kolb et al. 1994).

p. 222 **The PAH World:** [This section is based in part on extensive notes provided by Simon Nicholas Platts to RMH, 10 August 2004]

p. 223 **Max Bernstein:** Bernstein's seminar on November 19, 2001, was entitled "Interstellar Inception of Meteoritic Organics and Implications for the Origin of Life." The seminar is archived at www.origins.rpi.edu. The talk on ultraviolet irradiation of PAHs was based on research described in Bernstein et al. (1999b). Becker et al. (1997) present analytical evidence for PAHs delivered from space.

p. 223 **September 2003 conference:** The Seventh Trieste Conference on Chemical Evolution and the Origin of Life was held at the International Centre for Theoretical Physics, Trieste, Italy.

p. 224 **"On the return flight":** Nick Platts' airline ticket is dated September 30, 2003.

p. 224 **PAHs would have been abundant:** Desiraju and Gavezzotti (1989) present a useful review of PAH structures. An unfortunate confusion in nomenclature arises, because many chemists refer to PAHs as "polynuclear aromatics," or PNAs—the same abbreviation origin-of-life workers use for peptide nucleic acids.

p. 224 **"functionalized":** Bernstein et al. (1999b, 2003)

p. 229 **discotic organization:** Discotic self-organization was first described by Chandrasekhar et al. (1977). For reviews, see Chandrasekhar (1993), Kumar (2002, 2003), Chandrasekhar and Balagurusamy (2002), and Friedlein et al. (2003).

p. 230 **"I think it's worth pursuing..."**: [Jack Szostak to RMH, 3 June 2004]

p. 230 **"An experimental demonstration..."**: [Leslie Orgel to RMH, 4 June 2004]

p. 230 **"I thought it was interesting..."**: [Gerald Joyce to RMH, 7 June 2004]

p. 230 **"For now it is fascinating..."**: [Andrew H. Knoll to RMH, 7 June 2004]

p. 230 **"Thank you for submitting..."**: [Editors of *Nature* to Simon Nicholas Platts, 10 June 2004]. We later learned that, owing to an unusual glut of submissions, they never considered the scientific merits of the piece.

p. 231 **HBC for a thousand dollars:** The Norwegian chemical company Chiron AS offers 10 milligrams of hexabenzocoronene for $1,078 including shipping. Other possible gratis sources included Prof. Klaus Muellen at the Max Planck Institute for Polymer Research in Mainz, Germany, and Prof. Shigeru Ohshima and Dr. Minoru Takekawa at Toho University in Chiba, Japan.

p. 231 **informal talk:** The Carnegie talk was part of an irregular series by the Geobiology Discussion Group. The NASA video seminar was part of the Forum for Astrobiology Research.

p. 231 **thesis defense:** The thesis incorporated a range of projects related to origins chemistry, though the PAH World idea formed the center-piece.

p. 232   **Dave had also published:** Deamer (1997, p. 249). Deamer en-visioned individual functionalized PAHs incorporated sideways into mem-branes, where they could absorb near-ultraviolet and blue wavelengths and effectively act as photoelectric power sources.

18
THE EMERGENCE OF COMPETITION

p. 233   **"It is evident ...":** Sagan (1961, p. 177).

p. 233   *Origin of Species:* Darwin (1859).

p. 233   **continue unabated:** Scientists who trivialize these public con-cerns regarding the validity of scientific accounts of life's origin and evolu-tion do so at their peril. The theory of evolution, by which random molecular changes led through a chance process to the human species, can be seen as raising perplexing philosophical questions about the origin and meaning of human existence (see, e.g., National Academy of Sciences 1998, Gould 1999, K. R. Miller 1999, Fry 2000, Ruse 2000, Pennock 2002, Witham 2002, and Forrest and Gross 2004).

p. 235   **Competition among self-replicating cycles:** See, for example, Ycas (1955) and de Duve (2005).

p. 236   **Here's what Spiegelman:** Mills et al. (1967). Note that the rapid evolution of this system involved a cycling imposed by the experimenters. Recall that cycling is one of four factors that may promote emergent behav-iors (see Chapter 1).

p. 237   **"Our ultimate goal ...":** Zimmer (2004, p. 37).

p. 237   **yeast chromosomes:** Murray and Szostak (1983).

p. 237   **"Even then ...":** [Jack Szostak to RMH, 21 August 2004] He continues: "We tried to build an artificial chromosome in yeast by putting together all the pieces known at the time—centromeres, origins of replica-tion, telomeres, genes. But the construct we made didn't behave like a nor-mal chromosome, which led to the discovery that simple length was important. . . . I'm still following this principle. We think we understand 'life'—if we're right, we should be able to build a cell that acts like it's alive."

p. 237   **first series of ribozyme experiments:** Doudna and Szostak (1989), Doudna et al. (1991), Bartel and Szostak (1992), and Green and Szostak (1991).

Szostak writes: "[Jennifer Doudna] was actually the first person in my lab other than myself to start working with ribozymes. . . . The other students involved in that work were David Bartel and Rachel Green—also both remarkable success stories." [Jack Szostak to RMH, 21 August 2004]

p. 237   **Szostak's objective:** Ellington and Szostak (1990, 1992), Bartel and Szostak (1993), Szostak and Ellington (1993), Sassanfar and Szostak (1993), and C. Wilson and Szostak (1995). Early work is described in Joyce (1992). For a review, see Lorsch and Szostak (1996).

Recent experiments by Szostak's group focus on the evolution of an alternative TNA molecule—the simpler analog of RNA that incorporates the 4-carbon sugar threose. Szostak writes: "By doing in vitro evolutionary experiments with TNA sequences (something we're just now able to start) we can see if it's possible to evolve TNA molecules with specific binding or catalytic properties. If the answer is yes, then one could at least imagine TNA based life-forms." [Jack Szostak to RMH, 21 August 2004]

p. 238   **David Bartel:** Szostak et al. (2001).

p. 238   **Synthetic Life:** See, for example, Szostak et al. (2001), Gibbs (2004), and Rasmussen et al. (2004). Szostak writes, "What we're doing is really no different from what you or any other pre-biotic chemist is trying to do. We define possible reactions, see what works and what doesn't, and hope it all points to a plausible pathway. Our experiments happen to involve polymers and large aggregates of small molecules, instead of small molecule reactions, but the principle of testing by experiment is the same." [Jack Szostak to RMH, 21 August 2004]

p. 239   **Martin Hanczyc and Shelley Fujikawa:** Hanczyc et al. (2003).

p. 239   **encapsulated RNA strands:** David Deamer has demonstrated another simple mechanism for incorporating RNA into vesicles. When he dried a solution with lipid vesicles and RNA strands, the vesicles collapsed into thin layers of molecules with RNA interleaved. When water was added, the vesicles reformed with RNA trapped inside (Shew and Deamer 1983). See also Deamer (1997, pp. 254-255).

Swiss biochemist Pier Luigi Luisi demonstrated that lipid vesicles could "replicate" if larger vesicles were forced through a fine mesh (Luisi and Varela 1989, Luisi 2004).

p. 239   **To test their ideas:** Chen et al. (2004, p. 1476) conclude, "Darwinian evolution at the organismal level might therefore have emerged earlier than previously thought—at the level of a one-gene cell."

When asked if this model represents a metabolism-first or genetics-first origin, Szostak replied: "I don't think it's metabolism first, at least in the way that's normally thought of, but it's also more than the simplest genetics-first models since there is both the spatial and informational aspect to the system. . . . Of course, if there is a way of having [RNA nucleotides] generated inside the vesicle by some network of metabolic reactions, that would be great. In fact, if the network of metabolic reactions works by the uptake of small membrane permeable molecules . . . and generates larger impermeable molecules, then the same osmotic effects would operate and drive vesicle growth, potentially providing a selection for better metabolic networks." [Jack Szostak to RMH, 7 September 2004]

p. 239   **"If we can get . . ."**: Jack Szostak, as quoted in *Howard Hughes Medical Institute News*, September 2, 2004.

19
THREE SCENARIOS FOR THE ORIGIN OF LIFE

p. 241   **"Anyone who tells you . . ."**: Kauffman (1995, p. 31).

p. 243   **cooperative chemical phenomenon:** Such a symbiosis of metabolism and genetics would have foreshadowed events billions of years later, when precursors of our own complex cells are thought to have incorporated smaller, metabolically efficient cells. According to Margulis and Sagan (1995), mitochondria, the energy-processing organelles in eukaryotic cells, are symbionts.

EPILOGUE—
THE JOURNEY AHEAD

p. 245   **"Once to every man and nation . . ."**: American poet James Russell Lowell wrote this poem in 1845 for the *National Anti-Slavery Standard*. It subsequently became a rallying hymn for abolitionists.

p. 245   **emergence of consciousness:** The theory of emergence suggests that the path to self-aware life, if not specifically to the human species, may be deterministic. See, for example, Chaisson (2001) and Morowitz (2002).

Bibliography

Abelson, P. H. 1956. Amino acids formed in primitive atmospheres. *Science* 124:935.

Acevedo, O. L., and L. E. Orgel. 1986. Template-directed oligonucleotide ligation on hydroxylapatite. *Nature* 321:790-792.

Adams, M. W. W. 1992. Novel iron-sulfur centers in metalloenzymes and redox proteins from extremely thermophilic bacteria. *Advances in Inorganic Chemistry* 38:341-396.

Alberts, B. M. 1986. The function of hereditary materials: Biological catalyses reflect the cell's evolutionary history. *American Zoologist* 26:781-796.

Alexander, J. 1948. *Life, Its Nature and Origin.* New York: Reinhold.

Allamandola, L. J. and D. M. Hudgins. 2003. From interstellar polycyclic aromatic hydrocarbons and ice to astrobiology. In V. Pironello et al. (editors), *Solid State Astrobiology* (NATO ASI). Dordrecht: Kluwer, pp. 251-316.

Allamandola, L. J., A. G. G. M. Tielens, and J. R. Barker. 1985. Polycyclic aromatic hydrocarbons and the unidentified infrared emission bands: Auto exhaust along the Milky Way! *Astrophysical Journal Letters* 290:L25.

Allamandola, L. J., A. G. G. M. Tielens, and J. R. Barker. 1989. Interstellar polycyclic aromatic hydrocarbons: The infrared emission bands, the excitation-emission mechanism and the astrophysical implications. *Astrophysical Journal Supplement Series* 71:733-775.

Allamandola, L. J., S. A. Sandford, and M. P. Bernstein. 1997. Interstellar/precometary organic material and the photochemical evolution of complex organics. In C. B. Cosmovici, S. Bowyer, and D. Werthimer (editors), *Astrochemical and Biochemical Origins and the Search for Life in the Universe.* Bologna: Editrice Compositori, pp. 23-49.

Allwood, A. C., M. R. Walter, C. P. Marshall, and M. J. Kranendonk. 2004. Life at 3.4 Ga: Paleobiology and paleoenvironment of the stromatolitic Strelley Pool Chert, Pilbara Craton, Western Australia [abstract 197-4]. Geological Society of America Annual Meeting. Denver, CO, November 10, 2004.

Altman, S., M. F. Baer, M. Bartkiewicz, H. Gold, C. Guerrier-Takeda, et al. 1989. Catalysis by the RNA subunit of Rnase—a minireview. *Gene* 82:63-64.

Anders, E. 1989. Pre-biotic organic matter from comets and asteroids. *Nature* 342: 255-257.

Apel, C. L., M. Mautner, and D. Deamer. 2000. Self-assembled vesicles of monocarboxylic acids and alcohols: A model membrane system for early cellular life [abstract]. First Astrobiology Science Conference. NASA Ames Research Center, Moffett Field, California, p. 50.

Armstrong, K. 2000. *The Battle for God.* New York: Knopf.

Arrhenius, G., B. Gedulin, and S. Mojzsis. 1993. Phosphate in models for chemical evolution. In C. Ponnamperuma and J. Chela-Flores (editors), *Chemical Evolution: Origin of Life.* Hampton, VA: A. Deepak, pp. 25-50.

Awramik, S. M., J. W. Schopf, and M. Walter. 1983. Filamentous fossil bacteria from the Archean of Western Australia. *Precambrian Research* 20:357-374.

Bachmann, P. A., P. L. Luisi, and J. Lang. 1992. Autocatalytic self-replicating micelles as models for prebiotic structures. *Nature* 357:57-59.

Bada, J. L. 1972. Dating of fossil bones using the racemization of isoleucine. *Earth and Planetary Science Letters* 15:223-231.

Bada, J. L. 2004. How life began on Earth: A status report. *Earth and Planetary Science Letters* 226:1-15.

Bada, J. L., and A. Lazcano. 2002. Some like it hot, but not the first biomolecules. *Science* 296:1982-1983.

Bada, J. L., and A. Lazcano. 2003. Prebiotic soup—revisiting the Miller experiment. *Science* 300:745-746.

Bada, J. L., and R. A. Schroeder. 1972. Racemization of isoleucine in calcareous marine sediments: Kinetics and mechanism. *Earth and Planetary Science Letters* 15:1-11.

Bada, J. L., B. P. Luyendyk, and J. B. Maynard. 1970. Marine sediments: Dating by the racemization of amino acids. *Science* 170:730-732.

Bada, J. L., S. L. Miller, and M. Zhao. 1995. The stability of amino acids at submarine hydrothermal vent temperatures. *Origins of Life and Evolution of the Biosphere* 25:111-118.

Bagnold, R. A. 1941. *The Physics of Blown Sand and Desert Dunes.* London: Chapman and Hall.

Bagnold, R. A. 1988. *The Physics of Sediment Transport by Wind and Water.* New York: American Society of Civil Engineers.

Bailey, J., A. Chrysostomou, J. H. Hough, T. M. Gledhill, A. McCall, S. Clark, F. Menard, and M. Tamura. 1998. Circular polarization in star-formation regions: Implications for biomolecular homochirality. *Science* 281:672-674.

Bain, J. D., C. Y. Switzer, A. R. Chamberlin, and S. A. Benner. 1992. Ribosome-mediated incorporation of non-standard amino acids into a peptide through expansion of the genetic code. *Nature* 356:537-539.

Bak, P. 1996. *How Nature Works: The Science of Self-Organized Criticality.* New York: Copernicus Press.

Ball, P. 2004. Water, water everywhere? *Nature* 427:19-20.

Ban, N., P. Nissen, J. Hansen, P. B. Moore, and T. A. Steitz. 2000. The complete atomic structure of the large ribosomal subunit at 2.4 Å resolution. *Science* 289:905-920.

Bangham, A. D., M. M. Standish, and J. C. Watkins. 1965. Diffusion of univalent ions across the lamellae of swollen phospholipids. *Journal of Molecular Biology* 13: 238-252.

Barbrook, A. C., C. J. Howe, N. Blake, and P. Robinson. 1998. The phylogeny of *The Canterbury Tales*. *Nature* 394:839.

Baross, J. A., and J. Deming. 1995. Growth at high temperatures: Isolation and taxonomy, physiology, and ecology. In D. M. Carl (editor), *Deep-Sea Hydrothermal Vents*. Boca Raton, FL: CRC Press, pp. 169-217.

Baross, J. A., and S. E. Hoffman. 1985. Submarine hydrothermal vents and associated gradient environments as sites for the origin and evolution of life. *Origins of Life and Evolution of the Biosphere* 15:327-345.

Bartel, D. P., and J. W. Szostak. 1991. Template-directed primer extension catalyzed by the *Tetrahymena* ribozyme. *Molecular and Cellular Biology* 11:3390-3394.

Bartel, D. P., and J. W. Szostak. 1993. Isolation of new ribozymes from a large pool of random sequences. *Science* 261:1411-1418.

Bass, B. L., and T. R. Cech. 1984. Specific interaction between self-splicing RNA of *Tetrahymena* and its guanosine substrate: Implications for biological catalysis by RNA. *Nature* 308:820-826.

Becker, L., D. P. Glavin, and J. L. Bada. 1997. Polycyclic aromatic hydrocarbons (PAHs) in Antarctic Martian meteorites, carbonaceous chondrites, and polar ice. *Geochimica et Cosmochimica Acta* 61:475-481.

Becker, L., B. Popp, T. Rust, and J. L. Bada. 1999. The origin of organic matter in the Martian meteorite ALH84001. *Earth and Planetary Science Letters* 167:71-79.

Been, M. D., and T. R. Cech. 1988. RNA as an RNA polymerase: Net elongation of an RNA primer catalyzed by the *Tetrahymena* ribozyme. *Science* 239:1412-1416.

Behe, M. J. 1996. *Darwin's Black Box: The Biochemical Challenge to Evolution*. New York: Touchstone.

Beinert, H., R. H. Holm, and E. Münck. 1997. Iron-sulfur clusters: Nature's modular, multipurpose structures. *Science* 277:653-659.

Benner, S. A. 2002. What is a biosignature? What makes a planet habitable? Matching Darwinian chemistry to solar system habitats [abstract]. *Astrobiology* 2:433.

Benner, S. A., and D. Hutter. 2002. Phosphates, DNA, and the search for nonterrean life: A second generation model for genetic molecules. *Bioorganic Chemistry* 30:62-80.

Bernal, J. D. 1949. The physical basis of life. *Proceedings of the Royal Society* 62A: 537-558.

Bernal, J. D. 1951. *The Physical Basis of Life*. London: Routledge and Kegan Paul.

Bernal, J. D. 1967. *The Origin of Life*. London: Weidenfeld and Nicholson.

Bernstein, M. P., S. A. Sandford, and L. J. Allamandola. 1999a. Life's far-flung raw materials. *Scientific American* 281(1):42-49.

Bernstein, M. P., S. A. Sandford, L. J. Allamandola, J. S. Gillette, S. J. Clemett, and R. N. Zare. 1999b. UV irradiation of polycyclic aromatic hydrocarbons in ices: Production of alcohols, quinones, and ethers. *Science* 283:1135-1138.

Bernstein, M. P., J. P. Dworkin, S. A. Sandford, G. W. Cooper, and L. J. Allamandola. 2002. Racemic amino acids from the ultraviolet photolysis of interstellar ice analogues. *Nature* 416:401-403.

Bernstein, M. P., M. H. Moore, J. E. Elsila, S. A. Sandford, L. J. Allamandola, and R. N. Zare. 2003. Side group addition to the polycyclic aromatic hydrocarbon coronene by proton irradiation in cosmic ice analogs. *Astrophysical Journal* 582:L25-L29.

Biebricher, C. K., et al. 1993. Template-directed and template-free RNA synthesis. *Journal of Molecular Biology* 231:175-179.

Blank, J. G., G. H. Miller, M. J. Ahens, and R. E. Winans. 2001. Experimental shock chemistry of aqueous amino acid solutions and the cometary delivery of prebiotic compounds. *Origins of Life and Evolution of the Biosphere* 31:15-51.

Blöchl, E., M. Keller, G. Wächtershäuser, and K. O. Stetter. 1992. Reactions depending on iron sulfide and linking geochemistry with biochemistry. *Proceedings of the National Academy of Sciences USA* 89:8117-8120.

Bolli, M., R. Micura, and A. Eschenmoser. 1997. Pyranosil-RNA: Chiroselective self-assembly of base sequences by ligative oligomerization of tetranucleotide-2′,3′-cyclophosphates (with a commentary concerning the origin of biomolecular homochirality). *Chemical Biology* 4:309-320.

Bondi, H. 2004. Thomas Gold (1920-2004). *Nature* 430:415.

Bonner, W. A. 1991. The origin and amplification of biomolecular chirality. *Origins of Life and Evolution of the Biosphere* 21:59-111.

Bonner, W. A. 1995. Chirality and life. *Origins of Life and Evolution of the Biosphere* 25:175-190.

Bonner, W. A., P. R. Kavasmaneck, F. S. Martin, and J. J. Flores. 1975. Asymmetric adsorption by quartz: A model for the prebiotic origin of optical activity. *Origins of Life* 6:367-376.

Boston, P. J., M. V. Ivanov, and C. P. McKay. 1992. On the possibility of chemosynthetic ecosystems in subsurface habitats on Mars. *Icarus* 95:300-308.

Boyce, C. K., R. M. Hazen, and A. H. Knoll. 2001. Nondestructive, in situ, cellular-scale mapping of elemental abundances including organic carbon in permineralized fossils. *Proceedings of the National Academy of Sciences USA* 98:5970-5974.

Boyce, C. K., G. D. Cody, M. L. Fogel, R. M. Hazen, C. M. O'D. Alexander, and A. H. Knoll. 2003. Chemical evidence for cell wall lignification and the evolution of tracheids in early Devonian plants. *International Journal of Plant Science* 164:691-702.

Brandes, J. A., N. Z. Boctor, G. D. Cody, B. A. Cooper, R. M. Hazen, and H. S. Yoder, Jr. 1998. Abiotic nitrogen reduction on the early Earth. *Nature* 395:365-367.

Brandes, J. A., R. M. Hazen, H. S. Yoder, Jr., and G. D. Cody. 2000. Early pre- and post-biotic synthesis of alanine: An alternative to the Strecker synthesis. In G. A. Goodfriend, M. J. Collins, M. L. Fogel, S. A. Macko, and J. F. Wehmiller (editors), *Perspectives in Amino Acid and Protein Geochemistry.* Oxford, UK: Oxford University Press, pp. 41-47.

Brasier, M. D., O. R. Green, A. P. Jephcoat, A. K. Kleppe, M. J. Van Kranendonk, J. F. Lindsay, A. Steele, and N. V. Grassineau. 2002. Questioning the evidence for Earth's oldest fossils. *Nature* 416:76-81.

Brasier, M. D., O. R. Green, J. F. Lindsay, and A. Steele. 2004. Earth's oldest (approximately 3.5 Ga) fossils and the "early Eden" hypothesis: Questioning the evidence. *Origins of Life and Evolution of the Biosphere* 34:257-269.

Brocks, J. J., G. A. Logan, R. Buick, and R. E. Summons. 1999. Archean molecular fossils and the early rise of eukaryotes. *Science* 285:1033-1036.

Brown, A. 1999. Upwelling of hot gas. *American Scientist* 87:372.

Buick, R., B. Rasmussen, and B. Krapez. 1998. Archean oil: Evidence for extensive hydro-carbon generation and migration 2.5-3.5 Ga. *American Association of Petroleum Geologists Bulletin* 82:50-69.

Butlerow, A. 1861. Formation of monosaccharides from formaldehyde [in French]. *Comptes Rendu Hebdomadaires des Scéances de l'Académie des Sciences (Paris)* 53: 145-167.

Butterfield, N. J. 1990. Organic preservation of non-mineralizing organisms and the taphonomy of the Burgess Shale. *Paleobiology* 16:272-286.

Cady, S. L., J. D. Farmer, J. P. Grotzinger, J. W. Schopf, and A. Steele. 2003. Morphological biosignature and the search for life on Mars. *Astrobiology* 3:351-368.

Cairns-Smith, A. G. 1968. The origin of life and the nature of the primitive gene. *Journal of Theoretical Biology* 10:53-88.

Cairns-Smith, A. G. 1977. Takeover mechanisms and early biochemical evolution. *Biosystems* 9:105-109.

Cairns-Smith, A. G. 1982. *Genetic Takeover and the Mineral Origins of Life.* Cambridge, UK: Cambridge University Press.

Cairns-Smith, A. G. 1985a. *Seven Clues to the Origin of Life.* Cambridge, UK: Cambridge University Press.

Cairns-Smith, A. G, 1985b. The first organisms *Scientific American* 252(6):90-100.

Cairns-Smith, A. G. 1988. The chemistry of materials for artificial Darwinian systems. *International Reviews in Physical Chemistry* 7:209-250.

Cairns-Smith, A. G. 2001. The origin of life: Clays. In D. Baltimore, R. Dulbecco, F. Jacob, and R. Levi-Montalcini (editors), *Frontiers of Life.* New York: Academic Press, pp. 169-192.

Cairns-Smith, A. G., and H. Hartman. 1986. *Clay Minerals and the Origin of Life.* Cambridge, UK: Cambridge University Press.

Cairns-Smith, A. G., P. Ingram, and G. L. Walker. 1972. Formose production by minerals: Possible relevance to the origin of life. *Journal of Theoretical Biology* 35:601-604.

Cairns-Smith, A. G., A. J. Hall, and M. J. Russell. 1992. Mineral theories of the origin of life and an iron sulfide example. *Origins of Life and Evolution of the Biosphere* 22: 161-180.

Camazine, S., J.-L. Deneubourg, N. R. Franks, J. Sneyd, G. Theraulaz, and E. Bonabeau. 2001. *Self-Organization in Biological Systems.* Princeton, NJ: Princeton University Press.

Campbell, D. B., G. J. Black, L. M. Carter, and S. J. Ostro. 2003. Radar evidence for liquid surfaces on Titan. *Science* 302:431-434.

Cech, T. R. 1990. Self-splicing of group I introns. *Annual Reviews of Biochemistry* 59: 543-568.

Cech, T. R. 1993. The efficiency and versatility of catalytic RNA: Implications for an RNA world. *Gene* 135:33-36.

Cech, T. R. 2000. The ribosome is a ribozyme. *Science* 289:878-879.

Chaisson, E. J. 2001. *Cosmic Evolution: The Rise of Complexity in Nature.* Cambridge, MA: Harvard University Press.

Chandrasekhar, S. 1993. Discotic liquid crystals: A brief review. *Liquid Crystals* 14:3-14.

Chandrasekhar, S. and V. S. K. Balagurusamy. 2002. Discotic liquid crystals as quasi-one-dimensional electrical conductors. *Proceedings of the Royal Society of London,* Series A 458:1783-1794.

Chandrasekhar, S., B. K. Sadashiva, and K. A. Suresh. 1977. Liquid crystals of disk-like molecules. *Pramana Journal of Physics* 9:471-480.

Chapelle, F. H., K. O'Neill, P. M. Bradley, B. A. Methé, S. A. Ciufo, L. L. Knobel, and D. R. Lovley. 2002. A hydrogen-based subsurface microbial community dominated by methanogens. *Nature* 415:312-314.

Chen, I. A., R. W. Roberts, and J. W. Szostak. 2004. The emergence of competition between model protocells. *Science* 305:1474-1476.

Chyba, C. F., and C. Sagan. 1992. Endogenous production, exogenous delivery, and impact-shock synthesis of organic molecules: An inventory for the origins of life. *Nature* 355:125-132.

Chyba, C. F., P. J. Thomas, L. Brookshaw, and C. Sagan. 1990. Cometary delivery of organic molecules to the early Earth. *Science* 249:366-373.

Clark, S. 1999. Polarized starlight and the handedness of life. *American Scientist* 87: 336-343.

Clark, W. B. 1904. *Maryland Geological Survey: Miocene.* Baltimore, MD: The Johns Hopkins University Press. 2 volumes.

Cleland, C., and C. Chyba. 2002. Defining life. *Origins of Life and Evolution of the Biosphere* 32:387-393.

Cody, G. D. 2004. Transition metal sulfides and the origins of metabolism. *Annual Review of Earth and Planetary Sciences* 32:569-599.

Cody, G. D., N. Z. Boctor, T. R. Filley, R. M. Hazen, J. H. Scott, A. Sharma, and H. S. Yoder, Jr. 2000. Primordial carbonylated iron-sulfur compounds and the synthesis of pyruvate. *Science* 289:1337-1340.

Cody, G. D., C. M. O'D. Alexander, and F. Tera. 2001a. Solid-state (^{1}H and ^{13}C) nuclear magnetic resonance spectroscopy of insoluble organic residue in the Murchison meteorite: A self-consistent quantitative analysis. *Geochimica et Cosmochimica Acta* 66:1851-1865.

Cody, G. D., N. Z. Boctor, R. M. Hazen, J. A. Brandes, H. J. Morowitz, and H. S. Yoder, Jr. 2001b. Geochemical roots of autotrophic carbon fixation: Hydrothermal experiments in the system citric acid–H_2O–($\pm$FeS)–($\pm$NiS). *Geochimica et Cosmochimica Acta* 65:3557-3576.

Cody, G. D., N. Z. Boctor, J. A. Brandes, T. R. Filley, R. M. Hazen, and H. S. Yoder, Jr. 2004. Assaying the catalytic potential of transition metal sulfides for prebiotic carbon fixation. *Geochimica et Cosmochimica Acta* 68:2185-2196.

Colwell, F. S., T. C. Onstott, M. E. Dilwiche, D. Chandler, J. K. Frederickson, Q.-J. Yao, J. P. McKinley, D. R. Boone, R. Griffiths, T. J. Phelps, D. Ringelberg, D. C. White, L. LaFreniere, D. Bakwill, R. M. Lehman, J. Konisky, and P. E. Long. 1997. Microorganisms from deep, high temperature sandstones: Constraints on microbial colonization. *FEMS Microbiology Review* 20:425-435.

Conn, M. M., and J. Rebek, Jr. 1994. The design of self-replicating molecules. *Current Opinion in Structural Biology* 4:629-635.

Cooke, B. 2004. *The Gathering of Infidels: A Hundred Years of the Rationalist Press Association.* Amherst, NY: Prometheus Books.

Corliss, J. B., J. Dymond, L. I. Gordon, J. M. Edmond, R. P. von Herzen, et al. 1979. Submarine thermal springs on the Galapagos rift. *Science* 203:1073-1083.

Corliss, J. B., J. A. Baross, and S. E. Hoffman. 1981. An hypothesis concerning the relationship between submarine hot springs and the origin of life on Earth. In X. Le Pichon, J. Debyser, and F. Vine (editors), *Proceedings of the 26th International Geological Congress, Geology of the Oceans Symposium,* Paris 1980; *Oceanologica Acta* 4(supplement): 59-69.

Crick, F. H. C. 1968. The origin of the genetic code. *Journal of Molecular Biology* 38: 367-379.

Cronin, J. R. 1989. Origin of organic compounds in carbonaceous chondrites. *Advances in Space Research* 9(2):59-64.

Cronin, J. R., and S. Chang. 1993. Organic matter in meteorites: Molecular and isotopic analyses of the Murchison meteorite. In J. M. Greenberg et al. (editors), *The Chemistry of Life's Origins.* Dordrecht, The Netherlands: Kluwer Academic Press, pp. 209-258.

Cronin, J. R., and S. Pizzarello. 1983. Amino acids in meteorites. *Advances in Space Research* 3:5-18.

Dalton, R. 2002. Squaring up over ancient life. *Nature* 417:782-784.

Dalton, R. 2004. Fresh study questions oldest traces of life in Akilia rock. *Nature* 429:688.

Dana, E. S. 1958. *A Textbook of Mineralogy,* 4th edition. New York: John Wiley & Sons.

Darwin, C. 1859. *The Origin of Species.* London: John Murray.

Davankov, V. A. 1997. The nature of chiral recognition: Is it a three-point interaction? *Chirality* 9:99-102.

Davies, P. C. W. 1999. *The Fifth Miracle: The Search for the Origin and Meaning of Life.* New York: Touchstone.

Deamer, D. W. 1997. The first living systems: A bioenergetic perspective. *Microbiology and Molecular Biology Review* 61:239-261.

Deamer, D. W. 2003. Self-assembly of organic molecules and the origin of cellular life. *Reports of the National Center for Science Education* 23(May-August):20-33.

Deamer, D. W., and G. R. Fleischaker (editors). 1994. *Origin of Life: The Central Concepts.* London: Jones and Bartlett.

Deamer, D. W., and R. M. Pashley. 1989. Amphiphilic components of the Murchison carbonaceous chondrite: Surface properties and membrane formation. *Origins of Life and Evolution of the Biosphere* 19:21-38.

de Duve, C. 1991. *Blueprint for a Cell: The Nature and Origin of Life.* Burlington, NC: Neil Patterson.

de Duve, C. 1995a. *Vital Dust: Life as a Cosmic Imperative.* New York: Basic Books.

de Duve, C. 1995b. The beginnings of life on Earth. *American Scientist* 83:428-437.

de Duve, C. 1998. Constraints on the origin and evolution of life. *Proceedings of the American Philosophical Society* 142:525-532.

de Duve, C. 2005. The onset of selection. *Nature* 433:581-582.

de Duve, C., and S. L. Miller. 1991. Two-dimensional life? *Proceedings of the National Academy of Sciences USA* 88:10014-10017.

de Graaf, R. M., and A. W. Schwartz. 2000. Reduction and activation of phosphate on the primitive Earth. *Origins of Life and Evolution of the Biosphere* 30:405-410.

de Graaf, R. M., J. Visscher, and A. W. Schwartz. 1995. A plausibly prebiotic synthesis of phosphonic acids. *Nature* 378:474-477.

Delsemme, A. H. 1991. Nature and history of the organic compounds in comets: An astrophysical view. In R. L. Newburn, M. Neugebauer, and J. Rahe (editors), *Comets in the Post-Halley Era*. Dordrecht, The Netherlands: Kluwer Academic Press, pp. 377-428.

Dembski, W. A. 1999. *Intelligent Design: The Bridge Between Science & Theology*. Downers Grove, IL: InterVarsity Press.

Dembski, W. A. 2004. *The Design Revolution: Answering the Toughest Questions About Intelligent Design*. Downers Grove, IL: InterVarsity Press.

Desiraju, G. R., and A. Gavezzotti. 1989. Crystal structures of polynuclear aromatic hydrocarbons. Classification, rationalization and prediction from molecular structure. *Acta Crystallographica* B45:473-482.

D'Hondt, S., S. Rutherford, and A. J. Spivak. 2002. Metabolic activity of subsurface life in deep-sea sediments. *Science* 295:2067-2070.

D'Hondt, S., and 34 others. 2004. Distributions of microbial activities in deep subseafloor sediments. *Science* 306:2216-2221.

Dick, S. J., and J. E. Strick. 2004. *The Living Universe: NASA and the Development of Astrobiology*. New Brunswick, NJ: Rutgers University Press.

Diederichsen, U. 1996. Pairing properties of alanyl peptide nucleic acids containing an amino acid backbone with alternating configuration. *Angewandt Chemie International Edition English* 35:445-448.

Diederichsen, U., and H. W. Schmitt. 1998. β-homoalanyl PNAs: Synthesis and indication of higher ordered structures. *Angewandt Chemie International Edition English* 37: 302-305.

Dobson, C. M., G. B. Ellison, A. F. Tuck, and V. Vaida. 2000. Atmospheric aerosols as prebiotic chemical reactors. *Proceedings of the National Academy of Sciences USA* 97:11864-11868.

Dolman, C. E. 1975. Lazzaro Spallanzani. In C. C. Gillispie (editor), *Dictionary of Scientific Biography*, Vol. XII. New York: Scribners, pp. 553-567.

Donaldson, D. J., H. Tervahattu, A. F. Truck, and V. Vaida. 2004. Organic aerosols and the origin of life: An hypothesis. *Origins of Life and Evolution of the Biosphere* 34:57-67.

Donofrio, R. R. 2003. Siljan Ring findings remain unchanged since 1984. University of Oklahoma, Exploration and Development Geosciences Web site (http://www.edge.ou.edu/).

Doolittle, W. F. 2000. Uprooting the tree of life. *Scientific American* 282(2):90-95.

Doudna, J. A., and J. W. Szostak. 1989. RNA catalyzed synthesis of complementary strand RNA. *Nature* 339:519-522.

Doudna, J. A., S. Couture, and J. W. Szostak. 1991. A multi-unit ribozyme that is a catalyst of and a template for complementary strand RNA synthesis. *Science* 251:1605-1608.

Doukov, T. I., T. M. Iverson, J. Seravalli, S. W. Ragsdale, and C. L. Drennan. 2002. A Ni-Fe-Cu center in a bifunctional carbon monoxide dehydrogenase/acetyl-CoA synthase. *Science* 298:567-572.

Drobner, E., H. Huber, G. Wächtershäuser, D. Rose, and K. O. Stetter. 1990. Pyrite formation linked with hydrogen evolution under anaerobic conditions. *Nature* 346: 742-744.

Dworkin, J. P., D. W. Deamer, S. A. Sandford, and L. J. Allamandola. 2001. Self-assembling amphiphilic molecules: Synthesis in simulated interstellar/precometary ices. *Proceedings of the National Academy of Sciences USA* 98:815-819.

Dyson, F. 1985. *Origins of Life*. Cambridge, UK: Cambridge University Press. Also 1999 2nd edition.

Egholm, M., O. Burchardt, P. E. Nielsen, and R. H. Berg. 1992. Peptide nucleic acids (PNA). Oligonucleotide analogues with an achiral peptide backbone. *Journal of the American Chemical Society* 114:1897-1898.

Ehrenfreund, P., and S. B. Charnley. 2000. Organic molecules in the interstellar medium, comets, and meteorites. *Annual Review of Astronomy and Astrophysics* 38:427-483.

Eigen, M. 1971. Self-organization of matter and the evolution of biological macromolecules. *Naturwissenschaften* 58:465-523.

Eigen, M., and P. Schuster. 1977. The hypercycle: A principle of natural self-organization: Part A. Emergence of the hypercycle. *Naturwissenschaften* 64:541-565.

Eigen, M., and P. Schuster. 1978a. The hypercycle. A principle of natural self-organization. Part B. The abstract hypercycle. *Naturwissenschaften* 65:7-41.

Eigen, M., and P. Schuster. 1978b. The hypercycle. A principle of natural self-organization. Part C. The realistic hypercycle. *Naturwissenschaften* 65:341-369.

Eigen, M., and P. Schuster. 1979. *The Hypercycle: A Principle of Natural Self-Organization*. Berlin: Springer-Verlag.

Ellington, A. F. 1999. Molecular origins and the null hypothesis: Motifs from our maker? *Biological Bulletin* 196:315-317.

Ellington, A. E., and J. W. Szostak. 1990. In vitro selection of RNA molecules that bind specific ligands. *Nature* 346:818-822.

Ellington, A. E., and J. W. Szostak. 1992. Selection in vitro of single-stranded DNA molecules that fold into specific ligand binding sites. *Nature* 355:850-852.

Ellison, G. B., A. F. Tuck, and V. Vaida. 1999. Atmospheric processing of organic aerosols. *Journal of Geophysical Research* 104:11633-11641.

Engel, M. H., and S. A. Macko. 1997. Isotopic evidence for extraterrestrial non-racemic amino acids in the Murchison meteorite. *Nature* 296:837-840.

Ertem, G., and J. P. Ferris. 1996. Synthesis of RNA oligomers on heterogeneous templates. *Nature* 379:238-240.

Ertem, G., and J. P. Ferris. 1997. Template-directed synthesis using the heterogeneous templates produced by montmorillonite catalysis: A possible bridge between the prebiotic and RNA worlds. *Journal of the American Chemical Society* 119:7197-7201.

Eschenmoser, A. 1991. Warum pentose- und nicht hexose-Nucleinsäuren? *Nachrichten aus Chemie, Technik und Laboratorium* 39:795-806.

Eschenmoser, A. 1993. Hexose nucleic acids. *Pure and Applied Chemistry* 65:1179-1188.

Eschenmoser, A. 1994. Chemistry of potentially prebiological natural products. *Origins of Life and Evolution of the Biosphere* 24:389-423.

Eschenmoser, A. 1999. Chemical etiology of nucleic acid structure. *Science* 284:2118-2124.

Eschenmoser, A. 2004. The TNA-family of nucleic acid systems: Properties and prospects. *Origins of Life and Evolution of the Biosphere* 34:277-306.

Evans, M. C. W., B. B. Buchanan, and D. I. Arnon. 1966. A new ferredoxin-dependent carbon reduction cycle in a photosynthetic bacterium. *Proceedings of the National Academy of Sciences USA* 55:928-934.

Farley, J. 1977. *The Spontaneous Generation Controversy from Descartes to Oparin*. Baltimore, MD: The Johns Hopkins University Press.

Fedo, C. M., and M. J. Whitehouse. 2002. Metasomatic origin of quartz-pyroxene rock, Akilia, Greenland, and implications for Earth's earliest life. *Science* 296:1448-1452.

Ferris, J. P. 1993. Catalysis and prebiotic synthesis. *Origins of Life and Evolution of the Biosphere* 23: 307-315.

Ferris, J. P. 1999. Prebiotic synthesis on minerals: Bridging the prebiotic and RNA worlds. *Biology Bulletin* 196:311-314.

Ferris, J. P., and G. Ertem. 1992. Oligomerization of ribonucleotides on montmorillonite—reaction of the 5′-phosphorimidazolide of adenosine. *Science* 257:1387-1389.

Ferris, J. P., and G. Ertem. 1993. Montmorillonite catalysis of RNA oligomer formation in aqueous solution—a model for the prebiotic formation of RNA. *Journal of the American Chemical Society* 115:12270-12275.

Ferris, J. P., P. C. Joshi, E. H. Edelson, and J. G. Lawless. 1978. HCN: A plausible source of purine, pyrimidines and amino acids on the primitive Earth. *Journal of Molecular Evolution* 11:293-311.

Ferris, J. P., C.-H. Huang, and W. T. Hagan, Jr. 1988. Montmorillonite: A multifunctional mineral catalyst for the prebiological formation of phosphate esters. *Origins of Life and Evolution of the Biosphere* 18:121-133.

Ferris, J. P., A. R. Hill, Jr., R. Liu, and L. E. Orgel. 1996. Synthesis of long prebiotic oligomers on mineral surfaces. *Nature* 381:59-61.

Ford, B. J. 1985. *Single Lens: The Story of the Simple Microscope.* London: Heinemann.

Forrest, B., and P. R. Gross. 2004. *Creationism's Trojan Horse: The Wedge of Intelligent Design.* New York: Oxford University Press.

Forterre, P. 1996. A hot topic: The origin of hyperthermophiles. *Cell* 85:789-792.

Foustoukos, D. I., and W. E. Seyfried, Jr. 2004. Hydrocarbons in hydrothermal vent fluids: The role of chromium-bearing catalysts. *Science* 304:1002-1005.

Fox, S. W. 1956. Evolution of protein molecules and thermal synthesis of biochemical substances. *American Scientist* 44(October):347-359.

Fox, S. W. 1959. Origin of life. *Science* 130:1622-1623.

Fox, S. W. 1960. How did life begin? *Science* 133:200-208.

Fox, S. W. 1965. A theory of macromolecular and cellular origins. *Nature* 205:328-340.

Fox, S. W. 1968. Spontaneous generation, the origin of life, and self-assembly. *Current Modern Biology* 2:235-240.

Fox, S. W. 1980. Life from an orderly cosmos. *Naturwissenschaften* 67:576-581.

Fox, S. W. 1984. Protenoid experiments and evolutionary theory. In M. W. Ho and P. T. Saunders (editors), *Beyond Neo-Darwinism.* New York: Academic Press, pp. 15-60.

Fox, S. W. 1988. *The Emergence of Life: Darwinian Evolution from the Inside.* New York: Basic Books.

Fox, S. W., and K. Dose. 1977. *Molecular Evolution and the Origin of Life.* New York: Marcel Dekker.

Fox, S. W., and K. Harada. 1958. Thermal copolymerization of amino acids to a product resembling protein. *Science* 128:1214.

Franck, E. U. 1987. Fluids at high pressures and temperatures. *Journal of Chemical Thermodynamics* 19:225-242.

Franck, E. U., and H. Weingartner. 1999. Supercritical water. In T. Letcher (editor) *Chemical Thermodynamics, IUPAC Chemistry for the 21st Century Monograph.* Malden, MA: Blackwell Science, pp. 105-119.

Frederickson, J. K., and T. C. Onstott. 1996. Microbes deep inside the Earth. *Scientific American* 275(4):68-73.

Frederickson, J. K., B. N. Bjornstad, F. S. Colwell, L. Krumholz, R. M. Lehman, P. E. Long, J. P. McKinley, T. C. Onstott, T. J. Phelps, D. B. Ringelberg, J. M. Suflita, and D. C. White. 1997. Pore-size constraints on the activity and survival of subsurface bacteria in a late Cretaceous shale-sandstone sequence, northwestern New Mexico. *Geomicrobiology Journal* 14:183-202.

Freund, F., H. Kathrein, H. Wengeler, R. Knobel, and H. J. Heinen. 1980. Carbon in solid solution in forsterite—a key to the untraceable nature of reduced carbon in terrestrial and cosmogenic rocks. *Geochimica et Cosmochimica Acta* 44:1319-1333.

Freund, F., H. Wengeler, and E. Wappler. 1983. Thermal expansion anomaly and low-temperature internal friction in MgO. *High Temperatures—High Pressures* 15:357-358.

Freund, F., A. D. Gupta, and D. Kumar. 1999. Carboxylic and dicarboxylic acids extracted from crushed magnesium oxide single crystals. *Origins of Life and Evolution of the Biosphere* 29:489-509.

Freund, F., A. Staple, and J. Scoville. 2001. Organic protomolecule assembly in igneous minerals. *Proceedings of the National Academy of Sciences USA* 98:2142-2147.

Friedlein, R., X. Crispin, C. D. Simpson, M. D. Watson, F. Jackel, W. Osikowicz, S. Marciniak, M. P. de Jong, P Samori, S. K. M. Jonsson, M. Fahlman, K. Mullen, J. P. Rabe, and W. R. Salaneck. 2003. Electronic structure of highly ordered films of self-assembled graphitic nanocolumns. *Physical Review B* 68:(#195414)1-7.

Fry, I. 2000. *The Emergence of Life on Earth: A Historical and Scientific Overview.* New Brunswick, NJ: Rutgers University Press, pp. 318-320.

Furnes, H., N. R. Banerjee, K. Muehlenbachs, H. Staudigel, and M. de Wit. 2004. Early life recorded in Archean pillow lavas. *Science* 304:578-581.

Galison, P. 1987. *How Experiments End.* Chicago, IL: University of Chicago Press.

Gamow, G. 1954. Possible relation between deoxyribonucleic acid and protein structure. *Nature* 173:318.

Gedulin, B., and G. Arrhenius. 1994. Sources and geochemical evolution of RNA precursor molecules: The role of phosphate. In S. Bengtson (editor), *Early Life on Earth*, Nobel Symposium 84. New York: Columbia University Press, pp. 91-110.

Gee, H. 2002. That's life. *Nature* 416:28.

Geison, G. L. 1974. Louis Pasteur. In C. C. Gillispie (editor), *Dictionary of Scientific Biography*, Vol. X. New York: Scribners, pp. 350-416.

Gell-Mann, M. 1994. *The Quark and the Jaguar.* New York: W. H. Freeman.

Gell-Mann, M. 1995. What is complexity? *Complexity* 1:16-19.

Gell-Mann, M., and S. Lloyd. 2004. Effective complexity. In M. Gell-Mann and C. Tsallis (editors), *Nonextensive Entropy: Interdisciplinary Applications.* Oxford, UK: Oxford University Press, pp. 387-398.

Gell-Mann, M., and C. Tsallis (editors). 2004. *Nonextensive Entropy: Interdisciplinary Applications.* Oxford, UK: Oxford University Press.

Gibbs, W. W. 2004. Synthetic life. *Scientific American* 290(5):74-81.

Gilbert, W. 1986. The RNA world. *Nature* 319:618.

Gingerich, P. D., B. H. Smith, and E.L. Simons. 1990. Hind limbs of Eocene *Basilosaurus*: Evidence of feet in whales. *Science* 249:154-157.

Gingerich, P. D., S. M. Raza, M. Arif, M. Anwar, and X. Zhou. 1994. New whale from the Eocene of Pakistan and the origin of cetacean swimming. *Nature* 368:844-847.

Gish, D. T. 1985. *Evolution: The Challenge of the Fossil Record.* El Cajon, CA: Creation-Life Publishers.

Glavin, D. P., J. L. Bada, K. L. F. Brinton, and G. D. McDonald. 1999. Amino acids in the Martian meteorite Nakhla. *Proceedings of the National Academy of Sciences USA* 96:8835-8838.

Gogarten, J. P., R. D. Murphey, and L. Olendzenski. 1999. Horizontal gene transfer: Pitfalls and promises. *Biological Bulletin* 196:359-362.

Gogarten-Boekels, M., E. Hilario, and J. P. Gogarten. 1995. The effects of heavy bombardment on the early evolution—the emergence of three domains of life. *Origins of Life and Evolution of the Biosphere* 25:251-264.

Gold, T. 1977. Rethinking the origins of oil and gas. *Wall Street Journal,* 8 June 1977.

Gold, T. 1979. Terrestrial sources of carbon and earthquake outgassing. *Journal of Petroleum Geology* 1:3-19.

Gold, T. 1992. The deep hot biosphere. *Proceedings of the National Academy of Sciences USA* 89:6045-6049.

Gold, T. 1997. An unexpected habitat for life in the universe? *American Scientist* 85: 408-411.

Gold, T. 1999. *The Deep Hot Biosphere.* New York: Copernicus.

Gold, T., and S. Soter. 1980. The deep-earth gas hypothesis. *Scientific American* 242(6): 154-161.

Goldschmidt, V. M. 1952. Geochemical aspects of the origin of complex organic molecules on the Earth, as precursors to organic life. *New Biology* 12:97-105.

Goldsmith, D. 1997. *The Hunt for Life on Mars.* New York: Dutton.

Goodfriend, G. A. 1991. Patterns of racemization and epimerization of amino acids in land snail shells over the course of the Holocene. *Geochimica et Cosmochimica Acta* 55:293-302.

Goodfriend, G. A., and S. J. Gould. 1997. Paleontology and chronology of two evolutionary transitions by hybridization in the Bahamian land snail *Cerion. Science* 274:1894-1897.

Goodfriend, G. A., S. J. Gould, G. Carpintero, and M. G. Harasewych. 2003. The Holocene fossil record of *Cerion* land snails along eastern Long Island, Bahamas: Evidence for rapid faunal change [Abstract]. *Geological Society of America Abstracts with Programs,* Seattle, Washington, November 2-5, 2003.

Gould, S. J. 1989. *Wonderful Life: The Burgess Shale and the Nature of History.* New York: Norton.

Gould, S. J. 1999. *Rock of Ages: Science and Religion in the Fullness of Life.* New York: Ballantine Library of Contemporary Thought.

Gould, S. J. 2002. *The Structure of Evolutionary Theory.* Cambridge, MA: Belknap Press.

Grady, M. M. 2000. *Catalogue of Meteorites.* Cambridge, UK: Cambridge University Press.

Green, R., and J. W. Szostak. 1992. Selection of a ribozyme that functions as a superior template in a self-copying reaction. *Science* 258:1910-1915.

Griffith, C. A., T. Owens, T. R. Geballe, J. Rayner, and P. Rannou. 2003. Evidence for the exposure of water ice on Titan's surface. *Science* 300:628-631.

Guerrier-Takada, C., K. Gardiner, T. Marsh, N. Pace, and S. Altman. 1983. The RNA moity of ribonuclease P is the catalytic subunit of the enzyme. *Cell* 35:849-857.

Hagmann, M. 2002. Between a rock and a hard place. *Science* 295:2006-2007.

Haldane, J. B. S. 1929. The origin of life. *The Rationalist Annual* 1929:3-10.

Hanczyc, M. M., S. M. Fujikawa, and J. W. Szostak. 2003. Experimental models of primitive cellular compartments: Encapsulation, growth and division. *Science* 302: 618-622.

Hansen, J. L., M. van Hecke, A. Haaning, C. Ellegaard, K. H. Andersen, T. Bohr, and T. Sams. 2001. Instabilities in sand ripples. *Nature* 410:324.

Hare, P. E., and P. H. Abelson. 1968. Racemization of amino acids in fossil shells. *Carnegie Institution of Washington Year Book* 66:526-528.

Hare, P. E., and R. M. Mitterer. 1967. Nonprotein amino acids in fossil shells. *Carnegie Institution of Washington Year Book* 65:362-364.

Hare, P. E., and R. M. Mitterer. 1969. Laboratory simulation of amino-acid diagenesis in fossils. *Carnegie Institution of Washington Year Book* 67:205-208.

Harvey, R. P., and H. Y. McSween, Jr. 1996. A possible high-temperature origin for the carbonates in the Martian meteorite ALH84001. *Nature* 382:49-51.

Hayes, B. 2004. Ode to the code. *American Scientist* 92:494-498.

Haywood, A. 1985. *Creation and Evolution*. London: Triangle Books.

Hazen, R. M. 1993. *The New Alchemists: Breaking Through the Barriers of High Pressure Research*. New York: Random House.

Hazen, R. M. 1998. What was life's first energy source? *The Planetary Report* 18:16-17.

Hazen, R. M. 2001. Life's rocky start. *Scientific American* 284(4):62-71.

Hazen, R. M. 2002. Emergence and the origin of life. In G. Palyi, C. Zucchi, and L. Caglioti (editors), *Fundamentals of Life*. New York: Elsevier, pp. 277-286.

Hazen, R. M. 2004. Chiral crystal faces of common rock-forming minerals. In G. Palyi, C. Zucchi, and L. Caglioti (editors), *Progress in Biological Chirality*. Oxford, UK: Elsevier, Chapter 11, pp. 137-151.

Hazen, R. M., and D. S. Sholl. 2003. Chiral selection on inorganic crystalline surfaces. *Nature Materials* 2:367-374.

Hazen, R. M., and J. S. Trefil. 1991. *Science Matters: Achieving Scientific Literacy*. New York: Doubleday.

Hazen, R. M., T. Filley, and G. A. Goodfriend. 2001. Selective adsorption of L- and D-amino acids on calcite: Implications for biochemical homochirality. *Proccedings of the National Academy of Sciences USA* 98:5487-5490.

Hazen, R. M., A. Steele, J. Toporski, G. D. Cody, M. L. Fogel, and W. T. Huntress, Jr. 2002. Biosignatures and abiosignatures [abstract]. *Astrobiology* 2:512-513.

Head, J. N., H. J. Melosh, and B. A. Ivanov. 2002. Martian meteorite lunch: High-speed ejecta from small craters. *Science* 298:1752-1756.

Heinen, W., and A. M. Lauwers. 1996. Organic sulfur compounds resulting from interaction of iron sulfide, hydrogen sulfide and carbon dioxide in an aerobic aqueous environment. *Origins of Life and Evolution of the Biosphere* 26:131-150.

Hennet, R. J. C., N. G. Holm, and M. H. Engel. 1992. Abiotic synthesis of amino acids under hydrothermal conditions and the origin of life: A perpetual phenomenon? *Naturwissenschaften* 79:361-365.

Hill, A. R., and L. E. Orgel. 2002. Synthesis of adenine from HCN tetramer and ammonium formate. *Origins of Life and Evolution of the Biosphere* 32:99-102.

Hill, A. R., Jr., C. Böhler, and L. E. Orgel. 1998. Polymerization on the rocks: Negatively-charged alpha-amino acids. *Origins of Life and Evolution of the Biosphere* 28:235-242.

Hoang, Q. Q., F. Sicheri, A. J. Howard, and D. S. C. Yang. 2003. Bone recognition mechanism of porcine osteocalcin from crystal structure. *Nature* 425:977-980.

Hoffman, R. 2001. Thermophiles in Kamchatka. *American Scientist* 89:20-23.

Holland, H. D. 1984. *The Chemical Evolution of the Atmosphere and Oceans.* Princeton, NJ: Princeton University Press.

Holland, H. D. 1997. Evidence for life on Earth more than 3,850 million years ago. *Science* 275:38-39.

Holland, J. H. 1995. *Hidden Order.* Reading, MA: Helix Books.

Holland, J. H. 1998. *Emergence: From Chaos to Order.* Reading, MA: Helix Books.

Holm, N. G. 1992. Why are hydrothermal systems proposed as plausible environments for the origin of life? *Origins of Life and Evolution of the Biosphere* 22:5-14.

Holm, N. G., G. Ertem, and J. P. Ferris. 1993. The binding and reactions of nucleotides and polynucleotides on iron oxide hydroxide polymorphs. *Origins of Life and Evolution of the Biosphere* 23:195-215.

Horgan, J. 1996. *The End of Science: Facing the Limits of Knowledge in the Twilight of the Scientific Age.* Reading, MA: Helix Books.

Hough, L., and A. F. Rogers. 1956. Synthesis of amino acids from water, hydrogen, methane and ammonia. *Journal of Physiology* 132:28-30.

Huber, C., and G. Wächtershäuser. 1997. Activated acetic acid by carbon fixation on (Fe,Ni)S under primordial conditions. *Science* 276:245-247.

Huber, C., and G. Wächtershäuser. 1998. Peptides by activation of amino acids with CO on (Ni,Fe)S surfaces: Implications for the origin of life. *Science* 281:670-672.

Huber, C., W. Eisenreich, S. Hecht, and G. Wächtershäuser. 2003. A possible primordial peptide cycle. *Science* 301:938-940.

Hutter, D., and S. A. Benner. 2003. Expanding the genetic alphabet. Non-epimerizing nucleoside with the pyDNA hydrogen bonding pattern. *Journal of Organic Chemistry* 68:9839-9842.

Irvine, W. M. 1998. Extraterrestrial organic matter: A review. *Origins of Life and Evolution of the Biosphere* 28:365-383.

Johnson, S. 2001. *Emergence: The Connected Lives of Ants, Brains, Cities, and Software.* New York: Scribner.

Joyce, G. F. 1989. RNA evolution and the origins of life. *Nature* 338:217-224.

Joyce, G. F. 1991. The rise and fall of the RNA world. *New Biology* 3:399-407.

Joyce, G. F. 1992. Directed molecular evolution. *Scientific American* 267(6):90-97.

Joyce, G. F. 1994. Foreword. In D. W. Deamer and G. R. Fleischacker (editors), *Origins of Life: The Central Concepts.* Boston: Jones and Bartlett, pp. xi-xii.

Joyce, G. F. 1996. The role of catalytic RNA in the origin and early evolution of life on Earth. *Origins of Life and Evolution of the Biosphere* 26:233-234.

Joyce, G. F., and L. E. Orgel. 1993. Prospects for understanding the RNA world. In R. F. Gesteland and J. F. Atkins, (editors), *The RNA World.* Cold Spring Harbor, NY: Cold Spring Harbor Laboratory Press, pp. 1-25.

Joyce, G. F., A. W. Schwartz, S. L. Miller, and L. E. Orgel. 1987. The case for an ancestral genetic system involving simple analogues of the nucleotides. *Proceedings of the National Academy of Sciences USA* 84:4398-4402.

Karagounis, G., and G. Coumonlos 1938. A new method for resolving a racemic compound. *Nature* 142:162-163.

Kasting, J. F. 1990. Bolide impacts and the oxidation state of carbon in the Earth's early atmosphere. *Origins of Life and Evolution of the Biosphere* 20:199-231.

Kasting, J. F. 1993. Earth's early atmosphere. *Science* 259:920-926.

Kasting, J. F. 1994. Earth's early atmosphere: How reducing was it? *Origins of Life and Evolution of the Biosphere* 24:265-266.

Kasting, J. F. 2001. The rise of atmospheric oxygen. *Science* 293:819-820.

Kathrein, H., and F. Freund. 1983. DC conductivity of arc-fused MgO. *High Temperatures—High Pressures* 15:351-352.

Kauffman, S. A. 1993. *The Origins of Order: Self-Organization and Selection in Evolution.* New York: Oxford University Press.

Kauffman, S. A. 1995. *At Home in the Universe: The Search for Laws of Self-Organization and Complexity.* New York: Oxford University Press.

Kauffman, S. A. 1996. Self-replication: Even peptides do it. *Nature* 382:496-497.

Keller, M., E. Blöchl, G. Wächtershäuser, and K. O. Stetter. 1994. Formation of amide bonds without a condensation agent and implications for origin of life. *Nature* 368:836-838.

Keppler, H., M. Wiedenbeck, and S. S. Shcheka. 2003. Carbon solubility in olivine and the mode of carbon storage in the Earth's mantle. *Nature* 424:414-416.

Kerr, R. A. 1990. When a radical experiment goes bust. *Science* 247:1177-1179.

Kerr, R. A. 2002a. Deep life in the slow, slow lane. *Science* 296:1056-1058.

Kerr, R. A. 2002b. Reversals reveal pitfalls in spotting ancient and e.t. life. *Science* 296:1384-1385.

Kessler, M. A., and B. T. Werner. 2003. Self-organization of sorted patterned ground. *Science* 299:380-383.

Kidston, R., and W. H. Lang. 1917-1921. On Old Red Sandstone plants showing structure, from the Rhynie Chert bed, Aberdeenshire. *Transactions of the Royal Society of Edinburgh* 51:761-784; 52:603-627, 643-680; 52:855-902.

Knauth, L. P. 1998. Salinity history of the Earth's early ocean. *Nature* 395:554-555.

Knoll, A. H. 1996. Archean and Proterozoic paleontology. In J. Jansonius and D. C. McGegor, (editors), *Palynology: Principles and Applications,* Vol. 1. Dallas, TX: American Association of Stratigraphic Palynologists Foundation, Chapter 4, pp. 51-80.

Knoll, A. H. 1999. A new molecular window on early life. *Science* 285:1025-1026.

Knoll, A. H. 2003. *Life on a Young Planet: The First Three Billion Years of Evolution on Earth.* Princeton, NJ: Princeton University Press.

Koch, P. L. 1998. Isotopic reconstruction of past continental environments. *Annual Reviews of Earth and Planetary Sciences* 26:573-613.

Kolb, V. M., J. P. Dworkin, and S. L. Miller. 1994. Alternative bases in the RNA world—the prebiotic synthesis of urazole and its ribosides. *Journal of Molecular Evolution* 38:549-557.

Kornberg, A., and C. D. Fraley. 2000. Inorganic polyphosphate: A molecular fossil come to life. *ASM News* 66:275-280.

Kornberg, A., N. N. Rao, and D. Ault-Riché. 1999. Inorganic polyphosphate: A molecule of many functions. *Annual Review of Biochemistry* 68:89-125.

Kötz, J., F. Freund, and E. Klatt. 1983. Dielectric behavior of arc-fused MgO. *High Temperatures—High Pressures* 15:355-356.

Kruger, K., P. J. Grabowski, A. J. Zaug, J. Sands, D. E. Gottschling, and T. R. Cech. 1982. Self-splicing RNA: Autoexcision and autocyclization of the ribosomal RNA intervening sequence of *Tetrahymena*. *Cell* 31:147-157.

Krumholz, L. R., J. P. McKinley, Ulrich, G. A., and J. M. Suflita. 1997. Confined subsurface microbial communities in Cretaceous rock. *Nature* 386:64-66.

Kumar, S. 2002. Discotic liquid crystals for solar cells. *Current Science* 82:256-257.

Kumar, S. 2003. Molecular engineering of discotic nematic liquid crystals. *Pramana Journal of Physics* 61:199-203.

Kvenvolden, K. A., J. G. Lawless, K. Pering, E. Peterson, J. Flores, C. Ponnamperuna, I. R. Kaplan, and C. Moore. 1970. Evidence for extraterrestrial amino acids and hydrocarbons in the Murchison meteorite. *Nature* 28:923-928.

Kwok, S. 2004. The synthesis of organic and inorganic compounds in evolving stars. *Nature* 430:985-991.

Lahav, N. 1994. Minerals and the origin of life: Hypothesis and experiments in heterogeneous chemistry. *Heterogeneous Chemistry Review* 1:159-179.

Lahav, N. 1999. *Biogenesis*. Oxford, UK: Oxford University Press.

Lahav, N., and S. Chang. 1976. The possible role of solid surface area in condensation reactions during chemical evolution: Reevaluation. *Journal of Molecular Evolution* 8:357-380.

Lahav, N., D. White, and S. Chang. 1978. Peptide formation in the prebiotic era: Thermal condensation of glycine in fluctuating clay environments. *Science* 201:67-69.

Lasaga, A. C., H. D. Holland, and M. J. Dwyer. 1971. Primordial oil slick. *Science* 174:53-55.

Latora, V., and M. Marchiori. 2004. The architecture of complex systems. In M. Gell-Mann and C. Tsallis (editors), *Nonextensive Entropy: Interdisciplinary Applications*. Oxford: Oxford University Press, pp. 377-385.

Laudan, R. 1987. *From Mineralogy to Geology: The Foundations of a Science, 1650-1830*. Chicago, IL: University of Chicago Press.

Lazcano, A. 1986. Prebiotic evolution and the origin of cells. *Treballs de la Societat Catalana de Biologia* 39:73-103.

Lazcano, A., and S. L. Miller. 1994. How long did it take for life to begin? *Journal of Molecular Evolution* 39:546-554.

Lazcano, A., and S. L. Miller. 1996. The origin and early evolution of life: Prebiotic chemistry, the pre-RNA world, and time. *Cell* 85:793-798.

Lee, D. H., J. R. Granja, J. A. Martinez, K. Severin, and M. R. Ghadiri. 1996. A self-replicating peptide. *Nature* 382:525-528.

Lehman, N., and G. F. Joyce. 1993. Evolution in vitro of an RNA enzyme. *Nature* 361:182-185.

Lehninger, A. L., D. L. Nelson, and M. M. Cox. 1993. *Principles of Biochemistry*, 2nd edition. New York: Worth.

Lemke, K.H., D. S. Ross, J. L. Bischoff, R. J. Rosenbauern, and D. K. Bird. 2002. Hydrothermal stability of glycine and the formation of oligoglycine: Kinetics of peptide formation at 260°C and 200 bars [abstract]. 12th Annual V. M. Goldschmidt Conference, August 18-23, 2002, Davos, Switzerland.

Lepland, A., M. A. van Zuelin, G. Arrhenius, M. J. Whitehouse, and C. M. Fedo. 2005. Questioning the evidence for Earth's life—Akilia revisited. *Geology* 33:77-79.

Lerman, L. 1986. Potential role of bubbles and droplets in primordial and planetary chemistry exploration of the liquid-gas interface as a reaction zone for condensation process [abstract]. *Origins of Life and Evolution of the Biosphere* 16:201-202.

Lerman, L. 1994a. The bubble-aerosol-droplet cycle as a natural reactor for prebiotic organic chemistry (I) [abstract]. *Origins of Life and Evolution of the Biosphere* 24:111-112.

Lerman, L. 1994b. The bubble-aerosol-droplet cycle as a natural reactor for prebiotic organic chemistry (II) [abstract]. *Origins of Life and Evolution of the Biosphere* 24:138-139.

Lerman, L. 1996. The bubble-aerosol-droplet cycle: A prebiotic geochemical reactor [abstract]. *Origins of Life and Evolution of the Biosphere* 26:369-370.

Lévi-Strauss, C. 1969. *The Raw and the Cooked.* New York: Harper & Row

Lévi-Strauss, C. 1978. *Myth and Meaning: Cracking the Code of Culture.* Toronto, Canada: University of Toronto Press

Levin, G. V., and P. A. Straat. 1976. Viking labeled release biology experiment: Interim results. *Science* 194:1322-1329.

Levin, G. V., and P. A. Straat. 1981. A search for a non-biological explanation of the Viking labeled release life detection experiment. *Icarus* 45:494-516.

Li, T., and K. C. Nicolaou. 1994. Chemical self-replication of palindromic duplex DNA. *Nature* 369:218-221.

Lightman, A. 1993. *Einstein's Dreams: A Novel.* New York: Warner Books.

Lindsay, J. F., M. D. Brasier, N. McLoughlin, O. R. Green, M. Fogel, A. Steele, and S. Mertzman. 2003a. Archean hydrothermal systems and the nature of the abiotic/biotic boundary [abstract]. *EOS Transactions of the American Geophysical Union* 84:221.

Lindsay, J. F., M. D. Brasier, N. McLoughlin, O. R. Green, M. Fogel, K. M. McNamara, and A. Steele. 2003b. The Archean: Crucible of creation or cauldron of crud, data from Western Australia [abstract]. NASA Astrobiology Institute General Meeting 2003:110-111.

Liu, R., and L. E. Orgel. 1998. Polymerization on the rocks: Beta-amino acids and arginine. *Origins of Life and Evolution of the Biosphere* 28:245-257.

Löb, W. 1906. Studien über die chemische Wirkung de stillen elektrischen Entladung. *Zeitschrifte fur Elektrochemie* 11:282-316.

Löb, W. 1914. Über das Verhalten des Formamids unter der Wirkung der stillen Entladung. Ein Bertrag zur Frage Stickstoff-Assimilation. *Berichte der Deutschen Chemischen Gesellschaft* 46:684-697.

Lopez-Ruiz, R., H. L. Mancini, and X. Calbet. 1995. A statistical measure of complexity. *Physics Letters A* 209:321-326.

Lorenz, R. D., E. Kraal, E. Asphaug, and R. E. Thomson. 2003. The seas of Titan. *EOS Transactions of the American Geophysical Union* 84:125, 131-132.

Lorsch, J. R., and J. W. Szostak. 1996. Chance and necessity in the selection of nucleic acid catalysts. *Accounts of Chemical Research* 29:103-110.

Luisi, P. L. 1989. The chemical implementation of autopoesis. In G. R. Fleischaker, S. Colonna and P. L. Luisi (editors), *Self-Production of Supramolecular Structures*. Dordrecht, The Netherlands: Kluwer Academic Press, pp. 179-197.

Luisi, P. L. 2004. Introduction. *Origins of Life and Evolution of the Biosphere* 34:1-2.

Luisi, P. L., and F. J. Varela. 1989. Self-replicating micelles: A chemical version of a minimal autopoietic system. *Origins of Life and Evolution of the Biosphere* 19:633-643.

Luisi, P. L., P. Walde, and T. Oberholzer. 1994. Enzymatic RNA synthesis in self-reproducing vesicles: An approach to the construction of a minimum synthetic cell. *Berichte der Bunsengesellschaft für Physikalische Chemie* 98:1160-1165.

Lunine, J. I., Y. L. Yung, and R. D. Lorenz. 1999. On the volatile inventory of Titan from isotopic abundances in nitrogen and methane. *Planetary and Space Science* 47:1291-1303.

Madigan, M. T., and B. L. Marrs. 1997. Extremophiles. *Scientific American* 276(4):82-87.

Maher, K. A., and D. J. Stevenson. 1988. Impact frustration and the origin of life. *Nature* 331:612-614.

Margulis, L. 1999. The deep, hot biosphere (review). *Physics Today* (August):65.

Margulis, L., and D. Sagan. 1995. *What Is Life?* New York: Simon and Schuster.

Marshall, C. P., A. C. Allwood, M. R. Walter, M. J. Van Kranedonk, and R. E. Summons. 2004. Characterization of the carbonaceous material in the 3.4 Ga Strelley Pool Chert, Pilbara Craton, Western Australia [abstract 197-3]. *Geological Society of America Abstracts with Programs* 36:458.

Marshall, W. L. 1994. Hydrothermal synthesis of amino acids. *Geochimica et Cosmochimica Acta* 58:2099-2106.

Maynard-Smith, J., and E. Szathmáry. 1999. *The Origins of Life: From the Birth of Life to the Origin of Language*. New York: Oxford University Press.

McCollom, T. M., and J. S. Seewald. 2001. A reassessment of the potential for reduction of dissolved CO_2 to hydrocarbons during serpentinization of olivine. *Geochimica et Cosmochimica Acta* 65:3769-3778.

McCollom, T. M., and B. R. T. Simoneit. 1999. Abiotic formation of hydrocarbons and oxygenated compounds during thermal decomposition of iron oxalate. *Origins of Life and Evolution of the Biosphere* 29:167-186.

McKay, C. P., and W. J. Borucki. 1997. Organic synthesis in experimental impact shocks. *Science* 276:390-392.

McKay, D. S., E. K. Gibson, Jr., K. L. Thomas-Keprta, H. Vali, C. S. Romanek, S. J. Clemett, X. D. F. Chellier, C. R. Maechling, and R. N. Zare. 1996. Search for past life on Mars: Possible relic biogenic activity in Martian meteorite ALH84001. *Science* 273: 924-930.

McKay, D. S., S. Clemett, K. Thomas-Keprta, and E. K. Gibson. 2002. The classification of biosignatures [abstract]. *Astrobiology* 2:625-626.

McRae, C., C.-G. Gong, C. E. Snape, A. E. Fallick, and D. Taylor. 1999. $\delta^{13}C$ values of coal-derived PAHs from different processes and their application to source apportionment. *Organic Geochemistry* 30:881-889.

McSween, H. Y., Jr. 1994. What have we learned about Mars from SNC meteorites? *Meteoritics* 29:757-779.

Melosh, H. J. 1988. The rocky road to panspermia. *Nature* 332:687-688.

Melosh, H. J. 1993. Blasting rocks off planets. *Nature* 363:498-499.

Miller, K. R. 1999. *Finding Darwin's God: A Scientist's Search for Common Ground Between God and Evolution.* New York: Cliff Street Books.

Miller, S. L. 1953. A production of amino acids under possible primitive Earth conditions. *Science* 117:528-529.

Miller, S. L. 1955. Production of some organic compounds under possible primitive Earth conditions. *Journal of the American Chemical Society* 77:2351-2361.

Miller, S. L. 1997. Peptide nucleic acids and prebiotic chemistry. *Nature Structural and Molecular Biology* 4:167-169.

Miller, S. L., and J. L. Bada. 1988. Submarine hot springs and the origin of life. *Nature* 334:609-611.

Miller, S. L., and A. Lazcano. 2002. Formation of the building blocks of life. In J. W. Schopf (editor), *Life's Origin.* Berkeley, CA: University of California Press, pp. 78-112.

Miller, S. L., and H. C. Urey. 1959a. Organic compound synthesis on the primitive Earth. *Science* 130:245-251.

Miller, S. L., and H. C. Urey. 1959b. Reply. *Science* 130:1623-1624.

Mills, D. R., R. L. Peterson, and S. Spiegelman. 1967. An extracellular Darwinian experiment with a self-duplicating nucleic acid molecule. *Proceedings of the National Academy of Sciences USA* 58:217-224.

Mojzsis, S. J., G. Y. Fan, and G. Arrhenius. 1994. Phosphate microaggregates in Archean sediments [abstract]. *Origins of Life and Evolution of the Biosphere* 24:337.

Mojzsis, S. J., G. Arrhenius, K. D. McKeegan, T. M. Harrison, A. P. Nutman, and C. R. L. Friend. 1996. Evidence for life on Earth before 3,800 million years ago. *Nature* 384: 55-59.

Mojzsis, S. J., R. Krishnamurthy, and G. Arrhenius. 1998. After RNA and before: Geophysical and geochemical constraints on molecular evolution. In T. Cech et al. (editors), *The RNA World II.* Cold Spring Harbor, NY: Cold Spring Harbor Press, pp. 1-47.

Monastersky, R. 1997. Deep dwellers: Microbes thrive far below ground. *Science News* 151:192-193.

Monnard, P.-A., C. L. Apel, A. Kanavarioti, and D. W. Deamer. 2002. Influence of ionic inorganic solutes on self-assembly and polymerization processes related to early forms of life: Implications for a prebiotic aqueous medium. *Astrobiology* 2:139-152.

Monod, J. 1971. *Chance and Necessity.* Translated from the French by A. Wainhouse. New York: Knopf.

Moorbath, S., P. N. Taylor, and N. W. Jones. 1986. Dating the oldest terrestrial rocks—fact and fiction. *Chemical Geology* 57:63-86.

Morell, V. 1997. Microbiology's scarred revolutionary. *Science* 276:699-702.

Morowitz, H. J. 1992. *The Beginnings of Cellular Life: Metabolism Recapitulates Biogenesis.* New Haven, CT: Yale University Press.

Morowitz, H. J. 1996. Past life on Mars. *Science* 273:1639-1641.

Morowitz, H. J. 2002. *The Emergence of Everything.* New York: Oxford University Press.

Morowitz, H. J., B. Heinz, and D. W. Deamer. 1988. The chemical logic of a minimum protocell. *Origins of Life and Evolution of the Biosphere* 18:281-287.

Morowitz, H. J., J. D. Kostelnik, J. Yang, and G. D. Cody. 2000. The origins of intermediary metabolism. *Proceedings of the National Academy of Sciences USA* 97:7704-7708.

Morowitz, H. J., V. Srinivasan, and E. Smith. 2004. The paradigm shift in biochemistry. *Journal of the Washington Academy of Sciences* 90(3):58-66.

Muizon, C. de. 2001. Walking with whales. *Nature* 413:259-260.

Muñoz Caro, G. M., U. J. Meierhenrich, W. A. Schutte, B. Barbier, A. A. Segovia, H. Rosenbauer, W. H.-P. Thiemann, A. Brack, and J. M. Greenberg. 2002. Amino acids from ultraviolet irradiation of interstellar ice analogues. *Nature* 416:403-406.

Murray, A. W., and J. W. Szostak. 1983. The construction and properties of artificial chromosomes in yeast. *Nature* 305:189-193.

National Academy of Sciences. 1998. *Teaching About Evolution and the Nature of Science.* Washington, DC: National Academy Press.

National Academy of Sciences. 1999. *Science and Creationism: A View from the National Academy of Sciences,* 2nd edition. Washington, DC: National Academy Press.

Nealson, K. H. 1997a. Sediment bacteria: Who's there, what are they doing, and what's new? *Annual Review of Earth and Planetary Sciences* 25:403-434.

Nealson, K. H. 1997b. The limits of life on Earth and searching for life on Mars. *Journal of Geophysical Research* 102:23675-23686.

Nealson, K. H., and P. G. Conrad. 1999. Life: Past, present, and future. *Philosophical Transactions of the Royal Society of London,* Series B 354:1923-1939.

Nicolis, G., and I. Prigogine. 1989. *Exploring Complexity: An Introduction.* New York: W. H. Freeman.

Nielsen, P. E. 1993. Peptide nucleic acid (PNA): A model structure for the primordial genetic material? *Origins of Life and Evolution of the Biosphere* 23:323-327.

Nielsen, P. E. 1999. Peptide nucleic acid: A molecule with two identities. *Accounts of Chemical Research* 32:624-630.

Nielsen, P. E., M. Egholm, R. H. Berg, and O. Buchardt. 1991. Sequence-selective recognition of DNA by strand displacement with a thymine-substituted polyamide. *Science* 254:1497-1500.

Nielsen-Marsh, C. M., P. H. Ostrom, H. Ghandi, B. Shapiro, A. Cooper, P. V. Hauschka, and M. J. Collins. 2002. Sequence preservation of osteocalcin protein and mitochondrial DNA in bison bones older than 55 ka. *Geology* 30:1099-1102.

Nissen, P., J. Hansen, N. Ban, P. B. Moore, and T. A. Steitz. 2000. The structural basis of ribosome activity in peptide bond synthesis. *Science* 289:920-930.

Noller, H. F., V. Hoffarth, and L. Zimniak. 1992. Unusual resistance to protein extraction procedures. *Science* 256:1416-1419.

Nutman, A. P., V. R. McGregor, C. R. L. Friend, V. C. Bennett, and P. D. Kinny. 1996. The Itsaq Complex of southern West Greenland; the world's most extensive record of early crustal evolution (3900-3600 Ma). *Precambrian Research* 78:1-39.

Nutman, A. P., V. C. Bennett, C. R. L. Friend, and M. T. Rosing. 1997. ~3710 and >3790 Ma volcanic sequences in the Isua (Greenland) supracrustal belt; structural and Nd isotope implications. *Chemical Geology* 141:271-287.

Ohmoto, H. 1997. When did the Earth's atmosphere become oxic? *The Geochemical News* (93):12-13, 26-27.

Ohmoto, H., T. Kakegawa, and D. R. Lowe. 1993. 3.4-billion-year-old biogenic pyrite from Barberton, South Africa: Sulfur isotope evidence. *Science* 262:555-557.

Onstott, T. C., T. J. Phelps, F. S. Colwell, D. Ringelberg, D. C. White, D. R. Boone, J. P. Stevens, T. O. Stevens, D. L. Balkwill, T. Griffin, and T. Kleft. 1998. Observations pertaining to the origin and ecology of microorganisms recovered from the deep subsurface of Taylorsville Basin, Virginia. *Geomicrobiology Journal* 15:353-385.

Oparin, A. I. 1924. *Proiskhozhdenie Zhizni* (in Russian). Moscow: Moskovskii Rabochii. [Reprinted and translated by J. D. Bernal, 1967. *The Origin of Life*. London: Weidenfeld and Nicolson, pp. 199-234.]

Oparin, A. I. 1938. *The Origin of Life* (S. Morgulis, translator). New York: Macmillan.

Orgel, L. E. 1968. Evolution of the genetic apparatus. *Journal of Molecular Biology* 38: 381-393.

Orgel, L. E. 1986. RNA catalysis and the origin of life. *Journal of Theoretical Biology* 123:127-149.

Orgel, L. E. 1998a. The origin of life—a review of facts and speculations. *Trends in Biochemical Sciences* 23:491-495.

Orgel, L. E. 1998b. Polymerization on the rocks: Theoretical introduction. *Origins of Life and Evolution of the Biosphere* 28:227-234.

Orgel, L. E. 2000. A simpler nucleic acid. *Science* 290:1306-1307.

Orgel, L. E. 2003. Some consequences of the RNA World hypothesis. *Origins of Life and Evolution of the Biosphere* 33:211-218.

Oró, J. 1960. Synthesis of adenine from ammonium cyanide. *Biochemical and Biophysical Communications* 2:407-412.

Oró, J. 1961a. Mechanism of synthesis of adenine from hydrogen cyanide under possible primitive Earth conditions. *Nature* 191:1193-1194.

Oró, J. 1961b. Comets and the formation of biochemical compounds on the primitive Earth. *Nature* 190:389-390.

Orr, P. J., D. E. G. Briggs, and S. L. Kearns. 1998. Cambrian Burgess Shale animals replicated in clay minerals. *Science* 281:1173-1175.

Ourisson, G., and P. Albrecht. 1992. The hopanoids 1. Geohopanoids: The most abundant natural products on Earth? *Accounts of Chemical Research* 25:398-402.

Ourisson, G., and M. Rohmer. 1992. The hopanoids 2: The biohopanoids, a novel class of bacterial lipids. *Accounts of Chemical Research* 25:403-407.

Pace, N. R. 1997. A molecular view of microbial diversity and the biosphere. *Science* 276:734-740.

Pályi, G., C. Zucchi, and L. Caglioti (editors). 2002. *Fundamentals of Life*. New York: Elsevier.

Parkes, R. J. 1999. Oiling the wheels of controversy. *Nature* 401:644.

Parkes, R. J., B. A. Craig, S. J. Bale, J. M. Getiff, K. Goodman, P. A. Rochelle, J. C. Fry, A. J. Weightman, and S. M. Harvey. 1993. Deep bacterial biosphere in Pacific Ocean sediments. *Nature* 371:410-413.

Parsons, I., M. R. Lee, and J. V. Smith. 1998. Biochemical evolution II: Origin of life in tubular microstructures in weathered feldspar surfaces. *Proceedings of the National Academy of Sciences USA* 95:15173-15176.

Pasteris, J. D., and B. Wopenka. 2003. Necessary, but not sufficient: Raman identification of disordered carbon as a signature of ancient life. *Astrobiology* 3:727-738.

Pasteur, L. 1848. On the relations that can exist between crystalline form, chemical composition, and the sense of rotary polarization [original in French]. *Annales de Chimie Physique* 24:442-459.

Pedersen, K. 1993. The deep subterranean biosphere. *Earth-Science Reviews* 34:243-260.

Pedersen, K., S. Ekendahl, E. Tullborg, H. Furnes, I. Thorseth, and O. Tumyr. 1997. Evidence of life at 207 m depth in a granitic aquifer. *Geology* 25:827-830.

Pendleton, Y. J., and J. E. Chiar. 1997. The nature and evolution of interstellar organics. In Y. J. Pendleton and A. G. G. M. Tielens (editors), *From Stardust to Planetesimals*. ASP Conference Proceedings. 122:179-200.

Pennisi, E. 1998. Genome data shake tree of life. *Science* 280:672-674.

Pennock, R. T. (editor) 2002. *Intelligent Design Creationism and Its Critics: Philosophical, Theological, and Scientific Perspectives*. Cambridge, MA: MIT Press.

Peters, J. W. 2002. A trio of transition metals in anaerobic CO_2 fixation. *Science* 298: 552-553.

Peters, K. E., and M. J. Moldowan. 1993. *The Biomarker Guide*. Englewood Cliffs, NJ: Prentice Hall.

Peterson, E., F. Horz, and S. Chang. 1997. Modification of amino acids at shock pressures of 3.5 to 32 GPa. *Geochimica et Cosmochimica Acta* 61:3937-3950.

Piccirilli, J. A., T. Krauch, S. E. Moroney, and S. A. Benner. 1990. Extending the genetic alphabet. Enzymatic incorporation of a new base pair into DNA and RNA. *Nature* 343:33-37.

Pitsch, S., A. Eschenmoser, B. Gedulin, S. Hui, and G. Arrhenius. 1995. Mineral induced formation of sugar phosphates. *Origins of Life and Evolution of the Biosphere* 25: 297-334.

Pizzarello, S., and J. R. Cronin. 2000. Non-racemic amino acids in the Murray and Murchison meteorites. *Geochimica et Cosmochimica Acta* 64:329-338.

Plastino, A., M. T. Martin, and O. Rosso. 2004. Generalized information measures and the analysis of brain electrical signals. In M. Gell-Mann and C. Tsallis (editors), *Nonextensive Entropy: Interdisciplinary Applications*. Oxford, UK: Oxford University Press, pp. 261-293.

Podlech, J. 1999. New insight into the source of biomolecular homochirality: An extraterrestrial origin for molecules of life. *Angewandt Chemie International Edition English* 38:477-478.

Popa, R. 1997. A sequential scenario for the origin of biological chirality. *Journal of Molecular Evolution* 44:121-127.

Popper, K. R. 1959. *The Logic of Scientific Discovery*. London: Hutchinson.

Popper, K. R. 1963. *Conjectures and Refutations: The Growth of Scientific Knowledge*, 4th ed. London: Routledge and Kegan Paul.

Popper, K. R. 1972. *Objective Knowledge: An Evolutionary Approach*. London: Routledge.

Price, P. B. 2000. A habitat for psychrophiles in deep Antarctic ice. *Proceedings of the National Academy of Sciences USA* 97:1247-1251.

Prigogine, I. 1984. *Order out of Chaos: Man's New Dialogue with Nature*. Toronto: Bantam Books.

Radetsky, P. 1992. How did life start? *Discover* (November):74-82.

Radetsky, P. 1998. Life's crucible. *Earth* (February):34-41.

Rasmussen, S., L. Chen, D. Deamer, D. C. Krakauer, N. H. Packard, P. F. Stadler, and M. A. Bedau. 2004. Transitions from nonliving to living matter. *Science* 303:963-965.

Rawls, R. L. 2000. Iron-sulfur proteins. *Chemical and Engineering News* 79(Nov. 20): 43-51.

Rawls, R. L. 2002. Interstellar chemistry. *Chemical and Engineering News* 81(July 15): 31-37.

Rebek, J., Jr. 1994. Synthetic self-replicating molecules. *Scientific American* 271(1):48-55.

Rebek, J., Jr. 2002. Recognition, self-complementarity and autocatalysis. *Biokêmia* 26: 11-14.

Regis, E. 2003. *The Info Mesa*. New York: Norton.

Reid, C., and L. E. Orgel. 1967. Synthesis of sugars in potentially prebiotic conditions. *Nature* 216:455.

Reynolds, R. T., S. W. Squyres, D. S. Colburn, and D. S. McKay. 1983. On the habitability of Europa. *Icarus* 56:246-254.

Reysenbach, A.-L., S. Scitzinger, J. Kirshtein, and E. McLaughlin. 1999. Molecular constraints on a high-temperature evolution of early life. *Biological Bulletin* 196: 367-372.

Ricardo, A., M. A. Carrigan, A. N. Olcott, and S. A. Benner. 2004. Borate minerals stabilize ribose. *Science* 303:196-197.

Robertson, M. P., and S. L. Miller. 1995. An efficient prebiotic synthesis of cytosine and uracil. *Nature* 375:772-774.

Rohmer, M, P. Bouvier, and G. Ourisson. 1979. Molecular evolution of biomembranes: Structural equivalents and phylogenetic precursors of sterols. *Proceedings of the National Academy of Sciences USA* 76:847-851.

Rosing, M. T. 1999. ^{13}C-depleted carbon microparticles in >3700-Ma sea-floor sedimentary rocks from West Greenland. *Science* 283:674-676.

Ross, D. S., K. H. Lemke, and J. L. Bischoff. 2002a. The rapid exoergic oligomerization of amino acids [Abstract]. Abstracts with Program, Astrobiology Science Conference 2002 (p. 274), NASA Ames Research Center, Moffett Field, California.

Ross, D. S., K. H. Lemke, and J. L. Bischoff. 2002b. The strictly endogenous origins of amino acids: A chemical kinetics test [Abstract]. Abstracts with Program, Astrobiology Science Conference 2002 (p. 275), NASA Ames Research Center, Moffett Field, California.

Rudwick, M. J. S. 1976. *The Meaning of Fossils: Episodes in the History of Paleontology*, 2nd edition. Chicago, IL: University of Chicago Press.

Ruse, M. 2000. *Can a Darwinian Be a Christian? The Relationship Between Science and Religion*. Cambridge, UK: Cambridge University Press.

Russell, M. J., and A. J. Hall. 1997. The emergence of life from iron monosulphide bubbles at a submarine hydrothermal redox and pH front. *Journal of the Geological Society of London* 154:377-402.

Russell, M. J., and A. J. Hall. 2002. From geochemistry to biochemistry: Chemiosmotic coupling and transition element clusters in the onset of life and photosynthesis. *The Geochemical News* (113):6-12.

Russell, M. J., R. M. Daniel, and A. J. Hall. 1993. On the emergence of life via catalytic iron-sulfide membranes. *Terra Nova* 5:343-347.

Russell, M. J., R. M. Daniel, A. J. Hall, and J. Sherringham. 1994. A hydrothermally precipitated catalytic iron-sulfide membrane as a first step toward life. *Journal of Molecular Evolution* 39:231-243.

Sagan, C. 1961. On the origin and planetary distribution of life. *Radiation Research* 15: 174-192.

Sagan, C., W. R. Thompson, and B. N. Khare. 1992. Titan: A laboratory for prebiological organic chemistry. *Accounts of Chemical Research* 25:286-292.

Salam, A. 1991. The role of chirality in the origin of life. *Journal of Molecular Evolution* 33:105-113.

Sanchez, R., J. Ferris, and L. E. Orgel. 1966. Studies in prebiotic synthesis. *Science* 153: 72-73.

Sanchez, R., J. Ferris, and L. E. Orgel. 1967. Studies in prebiotic synthesis. II. Synthesis of purine precursors and amino acids from aqueous hydrogen cyanide. *Journal of Molecular Biology* 30:223-253.

Sassanfar, M., and J. W. Szostak. 1993. An RNA motif that binds ATP. *Nature* 364:550-553.

Schidlowski, M. 1988. A 3,800-million-year isotopic record of life from carbon in sedimentary rocks. *Nature* 333:313-318.

Schidlowski, M., J. M. Hayes, and I. R. Kaplan. 1983. Isotopic inferences of ancient biochemistries: Carbon, sulfur, hydrogen, and nitrogen. In J. W. Schopf (editor), *Earth's Earliest Biosphere: Its Origin and Evolution.* Princeton, NJ: Princeton University Press, pp. 149-186.

Schöning, K.-U., P. Scholz, S. Guntha, X. Wu, R. Krishnamurthy, and A. Eschenmoser. 2000. Chemical etiology of the nucleic acid structure: The α-threofuranosyl-$(3'\rightarrow2')$ oligonucleotide system. *Science* 290:1347-1351.

Schopf, J. W. (editor). 1983. *Earth's Earliest Biosphere: Its Origin and Evolution.* Princeton, NJ: Princeton University Press.

Schopf, J. W. 1993. Microfossils of the early Archean Apex Chert: New evidence of the antiquity of life. *Science* 260:640-646.

Schopf, J. W. 1999. *Cradle of Life: The Discovery of Earth's Earliest Fossils.* Princeton, NJ: Princeton University Press.

Schopf, J. W. (editor) 2002. *Life's Origin: The Beginnings of Biological Evolution.* Berkeley, CA: University of California Press.

Schopf, J. W., and B. M. Packer. 1987. Early Archean (3.3-billion to 3.5-billion-year-old) microfossils from Warrawoona Group, Australia. *Science* 237:70-73.

Schopf, J. W., and M. R. Walter. 1983. Archean microfossils: New evidence of ancient microbes. In J. W. Schopf (editor), *Earth's Earliest Biosphere.* Princeton, NJ: Princeton University Press, pp. 214-239.

Schopf, J. W., A. B. Kudryavtsev, D. G. Agresti, T. J. Wdowlak, and A. D. Czaja. 2002. Laser-Raman imagery of Earth's earliest fossils. *Nature* 416:73-76.

Schultz, S. 1999. Two miles underground. *Princeton Weekly Bulletin* 89(Dec. 13):1.

Schwartz, A. W., and R. M. de Graaf. 1993. The prebiotic synthesis of carbohydrates: A reassessment. *Journal of Molecular Evolution* 36:101-106.

Scott, E. R. D., A. Yamaguchi, and A. N. Krot. 1997. Petrological evidence for shock melting of carbonates in the Martian meteorite ALH84001. *Nature* 387:377-379.

Segré, D., D. Ben-Eli, D. W. Deamer, and D. Lancet. 2001. The lipid world. *Origins of Life and Evolution of the Biosphere* 31:119-145.

Shapiro, R. 1988. Prebiotic ribose synthesis: A critical analysis. *Origins of Life and Evolution of the Biosphere* 18:71-85.

Shapiro, R. 1995. The prebiotic role of adenine: A critical analysis. *Origins of Life and Evolution of the Biosphere* 25:83-98.

Shaw, R. W., T. B. Brill, A. A. Clifford, C. A. Eckert, and E. U. Franck. 1991. Supercritical water: A medium for chemistry. *Chemical and Engineering News* 69(51):26-39.

Sherwood-Lollar, B., S. K. Frape, S. M. Weise, P. Fritz, S. A. Macko, and J. A. Whelan. 1993. Abiogenic methanogenesis in crystalline rocks. *Geochimica et Cosmochimica Acta* 57:5087-5097.

Sherwood-Lollar, B., T. D. Westgate, J. A. Ward, G. F. Slater, and G. Lacrampe-Couloume. 2002. Abiogenic formation of alkanes in the Earth's crust as a minor source for global hydrocarbon reservoirs. *Nature* 416:522-524.

Shew, R., and D. Deamer. 1983. A novel method for encapsulating macromolecules in liposomes. *Biochimica Biophysica Acta* 816:1-8.

Shiner, J. S., M. Davison, and P. T. Landsberg. 1999. Simple measure for complexity. *Physical Review E* 59:1459-1464.

Shock, E. L. 1990a. Geochemical constraints on the origin of organic compounds in hydrothermal systems. *Origins of Life and Evolution of the Biosphere* 20:331-367.

Shock, E. L. 1990b. Do amino acids equilibrate in hydrothermal fluids? *Geochimica et Cosmochimica Acta* 54.1185-1189.

Shock, E. L. 1992a. Stability of peptides in high-temperature aqueous solutions. *Geochimica et Cosmochimica Acta* 56:3481-3491.

Shock, E. L. 1992b. Chemical environments of submarine hydrothermal systems. *Origins of Life and Evolution of the Biosphere* 22:67-108.

Shock, E. L. 1993. Hydrothermal dehydration of aqueous organic compounds. *Geochimica et Cosmochimca Acta* 57:3341-3349.

Shock, E. L., T. McCollum, and M. D. Schulte. 1995. Geochemical constraints on chemolithoautotrophic reactions in hydrothermal systems. *Origins of Life and Evolution of the Biosphere* 25:141-159.

Sievers, D., and G. von Kiedrowski. 1994. A self-replication of complementary nucleotide-based oligomers. *Nature* 369:221-224.

Simoneit, B. R. T. 1995. Evidence for organic synthesis in high temperature aqueous media—facts and prognosis. *Origins of Life and Evolution of the Biosphere* 25:119-140.

Simpson, S. 2003. Questioning the oldest signs of life. *Scientific American* 288(4):70-77.

Siveter, David J., M. D. Sutton, D. E. G. Briggs, and Derek J. Siveter. 2003. An ostracode crustacean with soft parts from the Lower Silurian. *Science* 302:1749-1751.

Sleep, N. H., K. Zahnle, J. F. Kasting, and H. J. Morowitz. 1989. Annihilation of ecosystems by large asteroid impacts on the early Earth. *Nature* 342:139-142.

Smith, E. and H. J. Morowitz. 2004. Searching for the laws of life. *SFI Bulletin* 19(Winter):16-23.

Smith, J. V. 1998. Biochemical evolution. I. Polymerization on internal, organophilic silica surfaces of dealuminated zeolites and feldspars. *Proceedings of the National Academy of Sciences USA* 95:3370-3375.

Smith, J. V., F. P. Arnold, Jr., I. Parsons, and M. P. Lee. 1999. Biochemical evolution. III: Polymerization on organophilic silica-rich surfaces, crystal-chemical modeling, formation of first cells, and geological clues. *Proceedings of the National Academy of Sciences USA* 96:3479-3485.

Sogin, M. L., et al. 1999. *Evolution: A Molecular Point of View.* Proceedings of a Workshop Sponsored by the Center for Advanced Studies in the Space Life Sciences at the MBL, 24-26 October 1997. Marine Biological Laboratory, Woods Hole, MA; *Biological Bulletin* 196:305-420.

Solé, R., and B. Goodwin. 2000. *Signs of Life: How Complexity Pervades Biology.* New York: Basic Books.

Sowerby, S. J., W. M. Heckl, and G. B. Petersen. 1996. Chiral symmetry breaking during the self assembly of monolayers from achiral purine molecules. *Journal of Molecular Evolution* 43:419-424.

Springsteen, G., and G. F. Joyce. 2004. Selective derivatization and sequestration of ribose from a prebiotic mix. *Journal of the American Chemical Society* 126:13-26.

Steele, A. J., D. T. Goddard, and I. B. Beech. 1994. An atomic force microscopy study of the biodeterioration of stainless steel in the presence of bacterial biofilms. *International Biodeterioration and Biodegredation* 33:35-46.

Steele, A., D. T. Goddard, D. Stapleton, J. K. Toporski, V. Peters, V. Bassinger, G. Sharples, D. D. Wynn-Williams, and D. S. McKay. 2000a. Investigations into an unknown organism on the Martian meteorite Allan Hills 84001. *Meteorite and Planetary Sciences* 35:237-241.

Steele, A., K. Thomas-Keprta, F. Westall, R. Avci, E. K. Gibson, C. Griffin, C. Whitby, D. S. McKay, and J. Toporski. 2000b. The microbiological contamination of meteorites: A null hypothesis [Abstract]. Abstracts with Program, First Astrobiology Science Conference, April 3-5, 2000 (p. 23). NASA Ames Research Center, Moffett Field, California.

Stetter, K. O., G. Fiala, G. Huber, R. Huber, and A Segerer. 1990. Hyperthermophilic microorganisms. *FEMS Microbiology Reviews* 75:117-124.

Stevens, T. O., and J. P. McKinley. 1995. Lithoautotrophic microbial ecosystems in deep basalt aquifers. *Science* 270:450-454.

Strick, J. E. 2000. *Sparks of Life: Darwinism and the Victorian Debate over Spontaneous Generation.* Cambridge, MA: Harvard University Press.

Sudarsan, N., J. E. Barrick, and R. R. Breaker. 2003. Metabolite-binding RNA domains are present in the genes of eukaryotes. *RNA* 9:644-647.

Summons, R. E., and M. R. Walter. 1990. Molecular fossils and microfossils of prokaryotes and protists from Proterozoic sediments. *American Journal of Science* 290A:212-244.

Summons, R. E., L. L. Jahnke, and B. R. T. Simoneit. 1996. Lipid biomarkers for bacterial ecosystems: studies of cultured organisms, hydrothermal environments and ancient sediments. In G. R. Bock and J. A. Goode (editors), *Evolution of Hydrothermal Systems on Earth (and Mars?).* New York: John Wiley & Sons, pp. 174-192.

Summons, R. E., L. L. Jahnke, J. M. Hope, and G. A. Logan. 1999. 2-Methylhopanoids as biomarkers for cyanobacterial oxygenic photosynthesis. *Nature* 400:554-557.

Switzer, C. Y., S. E. Moroney, and S. A., Benner. 1989. Enzymatic incorporation of a new base pair into DNA and RNA. *Journal of the American Chemical Society* 111:8322-8323.

Szostak, J. W., and A. D. Ellington. 1993. In vitro selection of functional RNA sequences. In R. F. Gesteland and J. F. Atkins (editors), *The RNA World.* Cold Spring Harbor, NY: Cold Spring Harbor Laboratory Press, pp. 511-533.

Szostak, J. W., D. P. Bartel, and P. L. Luisi. 2001. Synthesizing life. *Nature* 409:387-390.

Tanford, C. 1978. The hydrophobic effect and the organization of living matter. *Science* 200:1012-1018.

Thewissen, J. G. M., S. T. Hussain, and M. Arif. 1994. Fossil evidence for the origin of aquatic locomotion in archaeocete whales. *Science* 263:210-212.

Thewissen, J. G. M., E. M. Williams, L. J. Roe, and S. T. Hussain. 2001. Skeletons of terrestrial cetaceans and the relationship of whales to artiodactyls. *Nature* 413:277-281.

Thomas, D. N., and G. S. Dieckmann. 2002. Antarctic sea ice—a habitat for extremophiles. *Science* 295:641-644.

Thompson, W. R., and C. Sagan. 1992. Organic chemistry on Titan—surface interactions. *Proceedings of the Symposium on Titan, Toulouse, September 1991,* European Space Agency ESA SP-338, pp. 167-182.

Tian, F., O. B. Toon, A. A. Pavlov, and H. DeSterck. 2005. A hydrogen-rich early atmosphere. *Science* 308:1014-1017.

Tjivikua, T., Ballester, P., and Rebek, J. Jr. 1990. A self replicating system. *Journal of the American Chemical Society* 112:1249-1250.

Tödheide, K. 1972. Water at high temperatures and pressures. In F. Franks (editor), *Water: A Comprehensive Treatise: Vol. 1, The Physics and Physical Chemistry of Water.* New York: Plenum, pp. 463-514.

Toporski, J. K. W., A. Steele, F. Westall, R. Avci, D. M. Martill, and D. S. McKay. 2002. Morphological and spectral investigation of exceptionally well-preserved bacterial biofilms from the Oligocene Enspel formation, Germany. *Geochimica et Cosmochimica Acta* 66:1773-1791.

Trefil, J. S., and R. M. Hazen. 1992. *The Sciences: An Integrated Approach,* 1st ed. New York: John Wiley. Also 2nd edition (1999), 3rd edition (2001), and 4th edition (2004).

Trefil, J. S., and M. H. Hazen. 2002. *Good Seeing.* Washington, DC: Joseph Henry Press.

Tritz, J. P., D. Hermann, P. Bisseret, J. Connan, and M. Rohmer. 1999. Abiotic and biological hopanoid transformation: Towards the molecular fossils of the hopane series. *Organic Geochemistry* 30:499-514.

Tseng, H-Y., and T. C. Onstott. 1998. A tectogenic origin for the deep subsurface microorganisms of Taylorsville Basin: Thermal and fluid flow model constraints. *FEMS Microbiology Reviews* 20:391-397.

Tsuchida, R., M. Kobayashi, and A. Nakamura. 1935. Asymmetric adsorption of complex salts on quartz. *Journal of the Chemical Society of Japan* 56:1339.

Tuck, A. 2002. The role of atmospheric aerosols in the origin of life. *Surveys in Geophysics* 23:379-409.

Uchihashi, T., T. Okada, Y. Sugawara, K. Yokoyama, and S. Morita. 1999. Self-assembled monolayer of adenine base on graphite studied by noncontact atomic force microscopy. *Physical Review B* 60:8309-8313.

Uematsu, M., and E. U. Franck. 1981. Static dielectric constant of water and steam. *Journal of Physical and Chemical Reference Data* 9:1291-1306 (Note: volume is dated 1980).

Updike, J. 1986. *Roger's Version.* New York: Fawcett Crest.

Urey, H. C. 1951. The origin and develoment of the earth and other terrestrial planets. *Geochimica et Cosmochimica Acta* 1:209-277.

Urey, H. C. 1952. On the early chemical history of the Earth and the origin of life. *Proceedings of the National Academy of Sciences USA* 38:351-363.

Urey, H. C. 1966. Biological material in meteorites: A review. *Science* 151:157-166.

van Zuilen, M. A., A. Lepland, and G. Arrhenius. 2002. Reassessing the evidence for the earliest traces of life. *Nature* 418:627-630.

Von Baeyer, H. C. 1998. *Maxwell's Demon: Why Warmth Disperses and Time Passes.* New York: Random House.

Von Damm, K. L. 1999. Life on Earth. *Chemical and Engineering News* 78(Oct. 11): 123-124.

Von Kiedrowski, G. 1986. A self-replicating hexadeoxynucleotide. *Angewandt Chemie International Edition English* 25:932-935.

von Nägeli, C. 1884. *Mechanische-Physiologische Theorie der Abstammungslehre.* Munich/ Leipzig, Germany: Oldenbourg.

Wächtershäuser, G. 1988a. Pyrite formation, the first energy source for life: A hypothesis. *Systematic Applied Microbiology* 10:207-210.

Wächtershäuser, G. 1988b. Before enzymes and templates: Theory of surface metabolism. *Microbology Review* 52:452-484.

Wächtershäuser, G. 1990a. Evolution of the first metabolic cycles. *Proceedings of the National Academy of Sciences USA* 87:200-204.

Wächtershäuser, G. 1990b. The case for the chemoautotrophic origin of life in an iron-sulfur world. *Origins of Life and Evolution of the Biosphere* 20:173-176.

Wächtershäuser, G. 1991. Biomolecules: The origin of their optical activity. *Medical Hypotheses* 36:307-311.

Wächtershäuser, G. 1992. Groundworks for an evolutionary biochemistry: The iron-sulfur world. *Progress in Biophysics and Molecular Biology* 58:85-201.

Wächtershäuser, G. 1993. The cradle chemistry of life: On the origin of natural products in a pyrite-pulled chemoautotrophic origin of life. *Pure and Applied Chemistry* 65:1343-1348.

Wächtershäuser, G. 1994. Life in a ligand sphere. *Proceedings of the National Academy of Sciences USA* 92:4283-4287.

Wächtershäuser, G. 1997. The origin of life and its methodological challenge. *Journal of Theoretical Biology* 187:483-494.

Wächtershäuser, G. 2000. Life as we don't know it. *Science* 289:1307-1308.

Wald, G. 1954. The origin of life. *Scientific American* 191(2):44-53.

Waldrop, M. M. 1992. *Complexity: The Emerging Science at the Edge of Order and Chaos.* New York: Simon and Schuster.

Walker, J. C. G. 1986. Carbon dioxide on the early Earth. *Origins of Life and Evolution of the Biosphere* 16:117-127.

Walter, M. R. (editor). 1976. *Stromatolites.* Amsterdam: Elsevier.

Walter, M. R. 1983. Archean stromatolites: Evidence of the Earth's earliest benthos. In J. W. Schopf (editor), *Earth's Earliest Biosphere.* Princeton, NJ: Princeton University Press, pp. 187-213.

Walter, M. R., J. Bauld, and T. Brock. 1972. Siliceous algal and bacterial stromatolites in hot spring and geyser effluents of Yellowstone National Park. *Science* 178:402-405.

Watson, J. D., and F. H. Crick. 1953. A structure for deoxyribose nucleic acid. *Nature* 171:737-738.

Weber, A. 1982. Formation of pyrophosphate on hydroxyapatite with thioesters as condensing agents. *BioSystems* 15:183-189.

Weber, A. 1995. Prebiotic polymerization: Oxidative polymerization of 2,3-dimercapto-1-propanol on the surface of iron(III) hydroxide oxide. *Origins of Life and Evolution of the Biosphere* 25:53-60.

Westbrook, R. H. 1974. John Turberville Needham. In C. C. Gillispie (editor), *Dictionary of Scientific Biography*, Vol. X. New York: Scribners, pp. 9-11.

Westheimer, F. H. 1987. Why nature chose phosphates. *Science* 235:1173-1178.

Wharton, D. A. 2002. *Life at the Limits: Organisms in Extreme Environments.* Cambridge, UK: Cambridge University Press.

Whitehouse, M. 2000. Time constraints on when life began: The oldest record of life on Earth? *Newsletter of the Geochemical Society* (103):10-14.

Whitfield, J. 2004. It's life ... isn't it? *Nature* 430:288-290.

Wicken, J. S. 1987. *Evolution, Thermodynamics, and Information.* New York: Oxford University Press.

Wills, C., and J. L. Bada. 2000. *The Spark of Life: Darwin and the Primeval Soup.* Cambridge, MA: Perseus.

Wilson, C., and J. W. Szostak. 1995. In vitro evolution of a self-alkylating ribozyme. *Nature* 374:777-785.

Wilson, E. K. 1996. Earth had life earlier than previously thought. *Chemical and Engineering News* 74(Nov. 11):10.

Wilson, E. K. 1998. Go forth and multiply. *Chemical and Engineering News* 76(Dec. 7): 40-44.

Wilson, M. 2003. Where do carbon atoms reside within Earth's mantle? *Physics Today* 56(10):21-22.

Winkler, W., A. Nahvi, and R. R. Breaker. 2002. Thiamine derivatives bind messenger RNAs directly to regulate bacterial gene expression. *Nature* 419:952-956.

Winkler, W., A. Nahvi, A. Roth, J. A. Collins, and R. R. Breaker. 2004. Control of gene expression by a natural metabolite-responsive ribozyme. *Nature* 428:281-286.

Wintner, E. A., Conn, M. M., and Rebek, J., Jr. 1994. Self-replicating molecules: A second generation. *Journal of the American Chemical Society* 116:8877-8884.

Witham, L. A. 2002. *Where Darwin Meets the Bible: Creationists and Evolutionists in America.* New York: Oxford University Press.

Wittung, P., P. E. Nielsen, O. Buchardt, M. Egholm, and B. Norden. 1994. DNA-like double helix formed by peptide nucleic acid. *Nature* 368:561-563.

Woese, C. R. 1967. *The Genetic Code.* New York: Harper & Row.

Woese, C. R. 1978. A proposal concerning the origin of life on the planet Earth. *Journal of Molecular Evolution* 13:95-101.

Woese, C. R. 1987. Bacterial evolution. *Microbiology Review* 51:221-271.

Woese, C. R. 1998. The universal ancestor. *Proceedings of the National Academy of Sciences USA* 95:6854-6859.

Woese, C. R. 2000. Interpreting the universal phylogenetic tree. *Proceedings of the National Academy of Sciences USA* 97:8392-8396.

Woese, C. R. 2002. On the evolution of cells. *Proceedings of the National Academy of Sciences USA* 99:8742-8747.

Woese, C. R., and G. E. Fox. 1977. Phylogenetic structure of the prokaryotic domain: The primary kingdoms. *Proceedings of the National Academy of Sciences USA* 74:5088-5090.

Wöhler, F. 1828. On the artificial production of urea. *Annalen der Physik und Chemie,* 88:253-256.

Wolfenden, R., and M. J. Snider. 2001. The depth of chemical time and the power of enzymes as catalysts. *Accounts of Chemical Research* 34:938-945.

Wolfram, S. 2002. *A New Kind of Science.* Champaign, IL: Wolfram Media, Inc.

Yamagata, Y., H. Watanabe, M. Saitoh, and T. Namba. 1991. Volcanic production of polyphosphates and its relevance to prebiotic evolution. *Nature* 352:516-519.

Yao, S., Ghosh, I., Zutshi, R., and Chmielewski, J. 1997. A pH-modulated, self-replicating peptide. *Journal of the American Chemical Society* 119:10559-10560.

Ycas, M. 1955. A note on the origin of life. *Proceedings of the National Academy of Sciences USA* 41:714-716.

Yoder, H. S., Jr. 1950. High-low quartz inversion up to 10,000 bars. *Transactions of the American Geophysical Union* 31:821-835.

Yoder, H. S., Jr. 2004. *Centennial History of the Carnegie Institution of Washington. Volume III. The Geophysical Laboratory.* Cambridge, UK: Cambridge University Press.

Zaug, A. J., and T. R. Cech. 1986. The intervening sequence RNA of *Tetrahymenia* is an enzyme. *Science* 231:470-475.

Zettler, L. A. A., F. Gómez, E. Zettler, B. G. Keenan, R. Amils, and M. I. Sogin. 2002. Eukaryotic diversity in Spain's river of fire. *Nature* 417:137.

Zimmer, C. 1993. The first cell. *Discover* (Nov.):73-78.

Zimmer, C. 2004. What came before DNA? *Discover* (June):34-41.

Zubay, G., and T. Mui. 2001. Prebiotic synthesis of nucleotides. *Origins of Life and Evolution of the Biosphere* 31:87-102.

Index

S